Henry Mills Alden

A Study of Death

Henry Mills Alden

A Study of Death

ISBN/EAN: 9783337403119

Printed in Europe, USA, Canada, Australia, Japan

Cover: Foto ©berggeist007 / pixelio.de

More available books at **www.hansebooks.com**

BY

HENRY MILLS ALDEN

AUTHOR OF

" GOD IN HIS WORLD: AN INTERPRETATION "

NEW YORK

HARPER & BROTHERS PUBLISHERS

1895

TO MY BELOVED WIFE

OBIIT MAI VIII., MDCCCXCV

MY earliest written expression of intimate thought or cherished fancy was for your eyes only; it was my first approach to your maidenly heart, a mystical wooing, which neglected no resource, near or remote, for the enhancement of its charm, and so involved all other mystery in its own.

In you childhood has been inviolate, never losing its power of leading me by an unspoken invocation to a green field, ever kept fresh by a living fountain, where the Shepherd tends his flock. Now, through a body racked with pain and sadly broken, still shines this unbroken childhood, teaching me Love's deepest mystery.

It is fitting, then, that I should dedicate to you this book touching that mystery. It has been written in the shadow, but illumined by the brightness of an angel's face seen in the darkness, so that it has seemed easy and natural for me to find at the thorn's heart a secret and everlasting sweetness far surpassing that of the rose itself, which ceases in its own perfection.

Whether that angel we have seen shall, for my need and comfort and for your own longing, hold back his greatest gift, and leave you mine in the earthly ways we know and love, or shall hasten to make the heavenly surprise, the issue in either event will be a home-coming: if *here*, yet already the deeper secret will have been in part disclosed; and if *beyond*, that secret, fully known, will not betray the fondest hope of loving hearts. Love never denied Death, and Death will not deny Love.

H. M. A.

May 1, 1895.

PREFACE

DEATH and Evil, as considered in this work, are essentially one, and belong to Life not only in its manifestation but in its creative, or genetic, quality. Life, in its principle, is not good or evil, mortal or immortal; but as creative it becomes evil as well as good, and is immortal only as including mortality. This is also true of its creative transformations, in that series which we call its development. It is also, from the beginning, redemption as it is creation. Redemption is creative and creation is redemptive. The fountain is clear, and the stream clears itself.

This is our proposition. It is not new. It was St. Paul's theme. Always it is the spiritual intuition as distinguished from the strictly ethical view of life. James Hinton, writing thirty-five years ago, insisted upon the positive and radical character of evil; but he excluded sin from this view—a reservation which seems to us unnecessary and which St. Paul did not make. The present work had been practically completed when the four volumes of Mr. Hinton's privately printed MSS. were placed in my hands. Of these volumes, comprising altogether about three thousand royal octavo pages, I have been able to examine only the first. I have found in this so many remarkable resemblances to positions which I have taken that, although the divergencies of view are equally remarkable, I feel under an obligation (such as would have no force in the case of a published work) to allude to the fact. Mr. Hinton lays more stress than I have done upon alternativity in cosmic processes, more, however, with reference to polarisation and the vibratile character of all motion than to the meaning I have had in view in what I have designated as tropic reaction. My idea of the term " Limit " more nearly corresponds to his use of it, though the application is not the same. He thoroughly understood the value of the paradox. Mr. Hinton's treatise is not devoted to any particular theme; it is meant to represent the history of a mind in its workings toward an interpretation of universal life; and so many of his propositions are of a tentative character, being subsequently modified and sometimes reversed, that only a critical survey of the entire MSS. would yield the residuum of his thought. No one reading his writings can fail to be impressed by the originality and depth of his interpretation or to regret that his life was not spared long enough to enable him to organise his work into special theses upon the subjects treated. He wrote at a time when the Darwinian hypothesis had been but recently broached, yet he anticipated much that has since been the result of patient scientific research. His little volume, entitled "The Mystery of Pain," by which alone he is known to the general reading public, taken in connection with his unpublished writings, convinces me that no writer could have given to the world a work of such philosophic value as he might have prepared on the subject I have undertaken. After all, perhaps there has been no deeper insight shown or more subtle interpretation offered in this field than is to be found in Robert Browning's poetry.

Recent science abounds in suggestions of which I have availed most freely. Science discloses redemption in the realm of matter, and helps us to see death in birth and, in all development, the radical disturbance. The course of science itself is redemptive ; lost in its specialisations, its confinement seeks release, and an angel appears in its prison. Even the reptile followed to the end of its course is seen to take to itself wings for ascension. The bee, closely observed, is seen to inject into each cell of honey some poison from his sting which makes the sweetness wholesome —a venom inherent in the virtue.

In my restatement of cosmic specialisation, following the clues furnished by science, I have sought to emphasise the creative quality of Life in all its transformations and the homely sense of things in a living universe: to see that Genesis is Kinship.

In our reasoning, which must be imaginative, our path is through a series of analogues, which cease to be helpful and, indeed, mislead us if they are not themselves transformed in their translation from one order of existence to another. Each successive order in the series of creative transformations is a version or flexion, shown, in due course of the general movement, as a reversion. Then we see that from the first the entire movement is reversion—the turning always a returning—so that the universe reflects Godward. We find that this reversion is conspicuously apparent in the organic kingdom. It is triumphantly manifest in the Christ-life.

But Death and Evil are continued (whatever their transformation) into every new order—even into the kingdom of heaven, being therein lifted into their own heaven, where they are seen for what, in creation and redemption, they essentially are.

Faith boldly occupies the field of pessimism, finding therein its largest hope.

                              HENRY MILLS ALDEN.

CONTENTS

PROEM

THE DOVE AND THE SERPENT

PROEM

THE DOVE AND THE SERPENT

THE Dove flies, and the Serpent creeps. Yet is the Dove fond, while the Serpent is the emblem of wisdom.

Both were in Eden: the cooing, fluttering, winged spirit, loving to descend, companion - like, brooding, following; and the creeping thing which had glided into the sunshine of Paradise from the cold bosoms of those nurses of an older world—Pain and Darkness and Death—himself forgetting these in the warmth and green life of the Garden. And our first parents knew nought of these as yet unutterable mysteries, any more than they knew that their roses bloomed over a tomb; so that when all animate creatures came to Adam to be named, the meaning of this living allegory which passed before him was in great part hidden, and he saw no sharp line dividing the firmament below from the firmament above; rather he leaned toward the ground, as one does in a garden, seeing how quickly it was fashioned into the climbing trees, into the clean flowers, and into his own shapely frame. It was upon the ground he lay when that deep sleep fell upon him from which he woke to find his mate, lithe as the serpent, yet with the fluttering heart of the dove.

As the Dove, though winged for flight, ever descended, so the Serpent, though unable to wholly leave the ground, tried ever to lift himself therefrom, as if to escape some ancient bond. The cool nights revived and nourished his memories of an older time, wherein lay his subtile wisdom, but day by day his aspiring crest grew brighter. The life of Eden became for him oblivion, the light of the sun obscuring and confounding his reminiscence, even as for Adam and Eve this life was Illusion, the visible disguising the invisible, and pleasure veiling pain.

In Adam the culture of the ground maintained humility. He was held, moreover, in lowly content by the charm of the woman, who was to him like the earth grown human; and since she was the daughter of Sleep, her love seemed to him restful as the night. Her raven locks were like the mantle of darkness, and her voice had the laughter of streams that lapsed into unseen depths.

But Eve had something of the Serpent's unrest, as if she too had come from the Underworld, which she would fain forget, seeking liberation, urged by desire as deep as the abyss she had left behind her and nourished from roots unfathomly hidden—the roots of the Tree of Life. She thus came to have conversation with the Serpent.

In the lengthening days of Eden's one Summer these two were more and more completely enfolded in the Illusion of Light. It was under this spell that, dwelling upon the enticement of fruit good to look at and pleasant to the taste, the Serpent denied Death, and thought of Good as separate from Evil. "Ye shall not surely die, but shall be as the gods, knowing good and evil." So far, in his aspiring day-dream, had the Serpent fared from his old familiar haunts—so far from his old-world wisdom!

A surer omen would have come to Eve had she listened to the plaintive notes of the bewildered Dove that in his downward flutterings had begun to divine what the Serpent had come to forget, and to confess what he had come to deny.

For already was beginning to be felt "the season's difference," and the grave mystery, without which Paradise itself could not have been, was about to be unveiled, the background of the picture becoming its foreground. The fond hands plucking the rose had found the thorn. Evil was known as something by itself, apart from Good, and Eden was left behind, as one steps out of infancy.

From that hour have the eyes of the children of men been turned from the accursed earth, looking into the blue above, straining their vision for a glimpse of white-robed angels.

Yet it was the Serpent that was lifted up in the wilderness; and when he who "became sin for us" was being bruised in the heel by the old enemy, the Dove descended upon him at his baptism. He united the wisdom of the Serpent with the harmlessness of the Dove. Thus in him were bound together and reconciled the elements which in human thought had been put asunder. In him Evil is overcome of Good, as in him Death is swallowed up of Life; and with his eyes we see that the robes of angels are white because they have been washed in blood.

FIRST BOOK

TWO VISIONS OF DEATH

CHAPTER I

THE BODY OF DEATH

LIFE has gone. There is no next breath, no return of the pulse. No stillness is so blank and void of all suggestion. The sculptured marble, through the arrest of motion, becomes forever mobile; but here the interruption is final, fixed in a frozen calm. There is here no poetic cæsura, or pause between two strains of the same harmony. The way in which these feet have walked has come to a full stop; of the motions and uses peculiar to this organism as a means of human expression there is no continuance.

Finality.

This abrupt conclusion begets in us a dull astonishment, as if we were suddenly come against a blank wall, an unyielding, insurmountable barrier. The operations of Nature, the most obvious and the most impressive, being forever recurrent, cultivate in us the habit of expectation, so that we refuse to accept finality. Lulls there may be, dividing pauses, but no absolute conclusion. The thing which hath been is that which shall be, and having the same form and character. The same sun forever rises again, and whatever the change of conditions, this change is itself repeated in the uniform succession of seasons. The disappearance of the individual organism, after its brief cycle, we scarcely note, since through the succession of generations we are surrounded by the same forms in all their variety; it is taken to heart only when the ties of kinship or cherished companionship are broken. Then, the first shock having passed—the wonder that one so full of life has come into this blind silence—a great wave bears us backward : we remember, and every memory has its thorn of sharp regret; every thought of what has been is pierced by the arrows of sorrow, as a cloud by lightnings, breaking into a storm of tears, because that which has been can never be again.

Expectation is paralysed by this dull, unanswering silence. There is no response to our love or our grief; no future for our waiting. We are in no presence; it is the brutal fact of absence that stares us in the face. We may not say that the beloved sleeps, for where is this sleeper, who has so suddenly fled that it is left for us to close the eyes and compose the rigid limbs? Instead of relaxation, as of one weary and brought to rest, there is extreme rigor, as of one entering upon some mighty travail. But this darkness veils not sleep nor the free play of dreams; and from it there is no waking either to work or to weep.

This is the mere body of death, held out to us in its stark and glacial calm for a moment of tender care, which for it has no meaning—for our tribute of tears, to which it is insensible; for the ritual of The After-part our grief and faith, in which it can have of a Mystery. no part. It offers no illusion ; every door is shut. It is a mute and surd in any human harmony, a senseless contradiction, a brutal negation, an irrational conclusion. If it were even dormant, then might we await a transformation, like that of the chrysalis, or like that which happened to this very organism when it emerged from its antenatal sleep. There is indeed to be a change, but not like that. Instead of a new synthesis, wherein, through a dormant larval mystery, an organism climbs into an upper chamber of the House of Life, freshly apparelled for a daintier bridal-feast—instead of this increment of beauty and wonder, we shall see dissolution, a sinking analytic motion, whereby every complexion simulating the proper character and habit of a man shall be obliterated. In this dissolving view all psychical and even all physiological suggestions vanish, and are seen to be impertinent to such processes as belong exclusively to the inorganic kingdom. So alien to humanity is this change that it is offensive to human sensibility and noxious to human health; and our most pressing concern, after mourning over our dead, is that we may bury it out of our sight.

A primal instinct urges the animal into seclusion at the approach of death, and leads men to cover their faces or turn them to the wall, signifying that here beginneth a mystery not open to outward observation. From the beginning this was the soul's supreme confessional, wherein it repented itself of the world, forsaking all trodden ways, acknowledging their finality and its own utter weariness of them, and was shown the hidden thoroughfare leading to the Father's house.

The mystery has passed before its mere after-part arrests our notice. There is in our staring eyes no more than in those of the dead any speculation that will help us to its comprehension. The gravedigger's philosophy is as shallow and noisome as the work of his hands. All considerations based upon what we see, or think we see, of death are empty fallacies. Hamlet at Ophelia's grave is not more fantastic in considering "to what base uses we may return" than is Claudio when he shapes his fears :

> "Aye, but to die and go we know not where ;
> To lie in cold obstruction, and to rot ;
> This sensible warm motion to become
> A kneaded clod ; and the delighted spirit
> To bathe in fiery floods, or to reside
> In thrilling regions of rock-ribbed ice ;
> To be imprisoned in the viewless winds,
> And blown with restless violence round about
> The pendent world ; or to be worse than worst
> Of those that lawless and uncertain thoughts
> Imagine howling ! 'Tis too horrible !"

To the physicist death is but the exact payment of man's debt to Nature, through the return of so much matter and so much force to that general fund of matter and of force which, in the scientific view, remains in all permutations forever the same unchangeable quantity. But the scales of the chemist or his crucible touch not the real mystery any more nearly than does the gravedigger's spade. And for the most part those homilies wherewith we help out the funereal gloss that we have put upon death have the same open-eyed emptiness and fatuity. Only to the closed eyes is there the true vision.

CHAPTER II

THE MYSTICAL VISION

THE Angel of Death is the invisible Angel of Life. While the organism is alive as a human embodiment death is present, having the same human distinction as the life, from which it is inseparable, be- ing indeed the better half of living — its winged half, its rest and inspiration, its secret spring of elasticity and quickness.   Life came upon the wings of Death, and so departs.

The Unseen Angel

If we think of life apart from death our thought is partial, as if we would give flight to the arrow without bending the bow.   No living movement either begins or is completed save through death.   If the shuttle return not there is no web; and the texture of life is woven through this tropic movement.

It is a commonly accepted scientific truth that the continuance of life in any living thing depends upon death.   But there are two ways of expressing this truth : one, regarding merely the outward fact, as when we say that animal or vegetable tissue is renewed through decay ; the other, regarding the action and re- action proper to life itself, whereby it forever springs freshly from its source.   The latter form of expression is mystical, in the true meaning of that term.   We close our eyes to the outward appearance, in order that we may directly confront a mystery which is already past before there is any visible indication thereof. Though the imagination engaged in this mystical apprehension borrows its symbols or analogues from observation and experience, yet these symbols are spiritually regarded by looking at life on its living side and abstracted as far as possible from outward embodiment. We especially affect physiological analogues because, being derived from our experience, we may the more readily have the inward regard of them; and by passing from one physiological analogue to another, and from all these to those furnished by the processes of nature outside of our bodies, we come to an apprehension of the action and reaction proper to life itself as an idea independent of all its physical representations.

Thus we trace the rhythmic beating of the pulse to the systole and diastole of the heart, and we note a similar alternation in the contraction and relaxation of all our muscles. Breathing is alternately inspiration and expiration. Sensation itself is by beats, and falls into rhythm. There is no uninterrupted strain of either action or sensibility; a current or a contact is renewed, having been broken. In psychical operation there is the same alternate lapse and resurgence. Memory rises from the grave of oblivion. No holding can be maintained save through alternate release. Pulsation establishes circulation, and vital motions proceed through cycles, each one of which, however minute, has its tropic of Cancer and of Capricorn. Then there are the larger physiological cycles, like that wherein sleep is the alternation of waking. Passing

from the field of our direct experience to that of observation, we note similar alternations, as of day and night,
summer and winter, flood and ebb tide ; and science
discloses them at every turn, especially in its recent
consideration of the subtle forces of Nature, leading
us back of all visible motions to the pulsations of the
ether.

Mechanism does not escape this trope and rhapsody,
being indeed their most conspicuous illustration, since
its fundamental principle is that of leverage, whereby
there is libration or oscillation, as of a scale or a pen-
dulum, or circular motion as of a wheel.   In celestial
mechanism the material fulcrum disappears, and there
is the invisible centre of motion, of flight and return,
through tendencies which seem to balance each other,
giving the motion the orbital form.

In the nebular hypothesis Science has presented us
a view of the development of the universe from a nebulous expanse, to which, in its final dissolution, it must
return.   This immense pulsation is the grand cycle,
the tropics of which evade all human calculation.

Now all these analogues or phenomenal representations of tropic movement lead us to the apprehension
of the trope as proper to life itself ; they are the formal imaginations of an imageless truth.   The trope itself vanishes into its invisible ground, and we have no
definite expression of it save in its manifestation.

The insistence, however, upon a mystical apprehension is not foreign to science, which demands for
its own completeness an invisible world.   To account
for the communication of energy through cosmic space,
the physicist postulates as a medium the invisible

ether, the vortical motions of which have displaced what were formerly known as the ultimate atoms. It is but a step from the ethereal vibration to the pulsation of the Eternal Life. We say pulsation, still clinging to an image, to the visible skirts of our expression of what is in itself ineffable, even as the Prophet was placed in the cleft of a rock and so had the vision of a God who had passed by, whose face no man can see. We behold that movement of pulsing life which is manifest, which is in time and which measures time; the alternate movement, outwardly apparent to us in dissolution only, is a vanishing from our view into a field whither we may not follow with the terms pertinent to existence in space and time—the field of a measureless eternal life. We are at a loss for predicates, and resort to negations. But that concerning which our negation is—that is Being itself, the ground of existence and of persistence, of appearance and of reappearance.

In considering the action and reaction proper to life itself, we here dismiss from view all measured cycles, whose beginning and end are appreciably separate; our regard is confined to living moments, so fleet that their beginning and ending meet as in one point, which is seen to be at once the point of departure and of return. Thus we may speak of a man's life as included between his birth and his death, and, with reference to this physiological term, think of him as living and then as dead; but we may also consider him while living as yet every moment dying, and in this view death is clearly seen to be the inseparable companion of life, the way of return and so of continuance. This pulsation, forever a vanishing and a resurgence, so incalculably swift as to escape observation, is proper to life as life, does not begin with what we call birth nor end with what we call death (considering birth and death as terms applicable to an individual existence); it is forever beginning and forever ending. Thus to all manifest existence we apply the term Nature (*natura*), which means *forever being born;* and on its vanishing side it is *moritura*, or *forever dying.* Resurrection is thus a natural and perpetual miracle. The idea of life as transcending any individual embodiment is as germane to science as it is to faith.

Death, thus seen as essential, is lifted above its temporary and visible accidents. It is no longer associated with corruption, but rather with the sweet and wholesome freshness of life, being the way of its renewal. Sweeter than the honey which Absolution. Samson found in the lion's carcass is this everlasting sweetness of Death; and it is a mystery deeper than the strong man's riddle.

So is Death pure and clean, as is the dew that comes with the cool night when the sun has set; clean and white as the snow-flakes that betoken the absolution which Winter gives, shriving the earth of all her Summer wantonness and excess, when only the trees that yield balsam and aromatic fragrance remain green, breaking the box of precious ointment for burial.

In this view also is restored the kinship of Death with Sleep.

The state of the infant seems to be one of chronic

2

mysticism, since during the greater part of its days its
eyes are closed to the outer world.   Its
Sleep and
Death. larger familiarity is still with the invisible,
and it almost seems as if the Mothers of
Darkness were still withholding it as their nursling,
accomplishing for it some mighty work in their proper
realm, some such fiery baptism of infants as is frequent-
ly instanced in Greek mythology, tempering them for
earthly trials.   The infant must needs sleep while this
work is being done for it ; it has been sleeping since the
work began, from the foundation of the world, and the
old habit still clings about it and is not easily laid aside.

In that new field now open to the nascent organism
—a field of conscious effort directed toward outward
ends — there is exhaustion and expenditure.   There
must also be a special restoration, and this is given in
the regular and measured sleep of the adolescent and
adult organism, corresponding to its measured energy.
This later sleep differs from that of the infant in that
it is the relief from weariness, the winning back of a
spent force.   In the main—that is, in all unconscious
activities—the burden is still borne by an unseen power,
but there is also a burden and strain felt by the individual as in some way his own, appreciable in his consciousness and subject to his arbitrary determination--
a burden which he may voluntarily increase or di-
minish.   The loosening of the strain he does not thus
feel to be of his own ordering.   Sleep comes to him as
does the night whereto it seems to belong.   He may
resist it, but it will come, overtaking even the sentinel
at his post; or, again, he may court it with all dili-
gence and it shall fly away.

That which we have been considering as the death that is in every moment is a reaction proper to life itself, waking or sleeping, whereby it is renewed, sharing at once Time and Eternity—time as outward form and eternity as its essential quality. Sleep is a special relaxation, relieving a special strain. As daily we build with effort and design an elaborate superstructure above the living foundation, so must this edifice nightly be laid in ruins. Sleep is thus a disembarrassment, the unloading of a burden wherewith we have weighted ourselves. Here again we are brought into a kind of repentance and receive absolution. Sleep is forgiveness.

In some deeper sense sleep is one with death, and is proper and essential to life itself. Life forever sleeps beneath the masque of wakefulness, as it forever dies beneath the masque of phenomenal existence. The more of life, the more of death and the more of sleep. Wakefulness is but partial, and is associated more especially with age than with youth. Sleep, also, as we know it, is partial, not the inmost withdrawal to its chamber of eternal rest. For the recovery of man's strength life gives him this partial release. A saving hand is stretched forth out of the darkness, snatching him from the world and locking his energies in suspense. The world of conscious experience is cut off by a temporarily impassable chasm, as if for the sleeper it had no existence; and yet it is only the desire for that world which is being renewed in this darkness.

That which we commonly call the dream, whose stuff is borrowed from the daylight, occurs only on the outskirts of the domain of sleep. It has been fancied

that in a deeper dream, never registered in conscious
memory, there may be a return to the associations of
former lives, but this deeper dream—if such a dream
may be—imageless and having no outward moorings,
must also be inhospitable to reminiscences of any previous individual existence.   Though there is a suspension of individual activity, there is still the confinement
of individuality itself, whose integrity is never disturbed
in any normal condition of life.   In hypnotism and insanity there may be a schism or refraction of the individual self, and even, it may be, the resumption of an
ancient habit and familiarity—an atavistic reversion—
but not in sleep.   Hypnotism seems to be a kind of
necromancy, whereby the hidden depths of consciousness are brought to the surface at the bidding of outward suggestion.   But in normal sleep, whatever response there may be to outward suggestion, there is no
displacement of " the abysmal deeps of personality."

Sleep, in this special sense, is, indeed, akin to Death.
But he stands this side of the veil, only simulating the
offices of his invisible brother, who stands at the very
font of Life, the hierophant of the Greater Mysteries—
those of the eternal life.   Death calls with the voice of
Life, calls from the central source to the remotest cir-
cumference of the universal life, calls with every pulsation of that life, and is, indeed, if we may use such an
image, the return beat of the pulse of the All-Father's
heart, the attraction of all being to its centre of rest in
that Father's bosom, whatever may be its separate
movements in the cycles of Time and Space.   Sleep is
the hierophant of a Minor Mystery, folding us in his
mantle of darkness, renewing the world's desire, recov-

ering Time. Death from within the veil instantaneously and every instant transforms life from its very source, recovering Eternity. Sleep is re-creation. Death is the mighty Negation, whereby all worlds vanish into that Nothing from which all worlds are made, the vast inbreathing of the Spirit of God for His ever repeated fiat of Creation. Sleep suspends the individuality within its embodiment. Death shows the inmost personality in a divine presence — that angel of each one of us which forever beholds the face of the Father.

, .

Our usual regard of death is one which brings into the foreground its accidental aspects, not pertinent to its essential reality. Even our grief for dear ones taken from us dwells upon our loss, upon the difference to us which death has made, and so our attention is diverted from the transcendent office. On the hither side Death has no true interpreter, and none returns from its true domain to be the witness of its invisible glory, none save the risen Lord. But though the loved ones gone cannot return to us, we shall go *Ascendent Ministration of Death* to them; and this faith which follows that which has vanished, the Christian hope of resurrection, lifts us to a point of vision from which it is possible for us to see death for what it really is as invisibly an ascending ministrant, whatever frailty and decrepitude may attend the visible descent.

The pagan idea of immortality insisted upon deathlessness. The Christian faith in resurrection gives death back to life as essential to its transformation. Death is swallowed up of Life included therein As

"Children of the Resurrection," we have no part in what is commonly called death—that visible declension and dissolution from which our life is withdrawn, together with our true death, leaving the grave no victory.

We have only to allow ourselves the liberty which science takes, to arrive at this view as a philosophical conviction. We have, indeed, in juvenescence a visible illustration of an ascent of life upon the hidden wings of death. If man were distinguished from all other organisms by the possession of perpetual youth, we who are accustomed to associate death only with decline might pronounce him deathless, limiting the province of mortality to those organisms whose descent maintains his levitation. Gravitation, which is the physical symbol of death, was before Newton not suspected as a cosmic principle. Things were seen to fall upon the earth, but the earth was not seen to fall toward the sun; there was, indeed, no appreciable evidence of such a tendency. Yet, wholly apart from such visible signs thereof, Newton's mystical imagination leaped to the truth (afterward reasonably confirmed) that all bodies are falling bodies; and in his expression of this truth he made gravitation something more than is indicated in the outward aspects of falling and weight—he called it an attraction, so that his thought became the mystical apprehension of an unseen but universal cosmic bond. Thus though man had never shown any visible signs of decline, some Newton would have arisen in the physiological field and asserted his mortality, see-

A Physical Analogue.

ing that in youth death is swallowed up of life, as gravitation is in the ascent of every organism and in the
sustained distance from the sun of every planet.

Every organism has an action and reaction quite distinct from those of inorganic substances, and which
vanish from our view before there is left behind merely
"the dust that riseth up and is lightly laid again." In
the complex human life there is much more that vanishes—the passing of a spiritual as well as a physiological mystery, far withdrawn from outward observation
before the sceptical physicist or pessimist seizes upon
the mere residuum or precipitate as the object of his
fruitless investigation—fruitless, at least, as having any
pertinence to human destiny. The body which Death
leaves behind is surrendered to that inorganic chemistry which was formerly in alliance with the more subtle
actions and reactions of a distinctively human life, and
to the physical bond of gravitation which was once the
condition of its consistency but which now brings it to
the dust. Are we any more mystical than Newton and
Laplace in our conviction that Death as a part of the
higher life is its unseen bond—the way of return to its
source?

In the cycle of every living organism there is a descending as well as an ascending movement—age as
well as youth, so that the forces to which
the outward structure is finally abandoned The Secret of
the Kuin.
seem to have upon it a lien anticipating
their full possession. This is simply saying that the
life and death proper to the organism are gradually
withdrawing before they together wholly vanish, leav-

ing the field to lower life and death. But there is no claim of the lower upon the higher, save through the surrender made by the higher as a part of its proper destiny. The signal of retreat is not given from without but from the inmost chamber of the citadel, where reside the will and intelligence which determined the distinctive architecture of the marvellous superstructure, and which hold also the secret of its ruin. That secret is itself genetic: invisibly it looks toward *palingenesis*—toward the higher transformation of the vanishing life, and visibly toward the outward succession of a new generation.

So Death is Janus-faced: toward an unseen resurrection, a reascendent ministration, and toward the visible resurgence of new life upon the earth, to which it ministers by descent and which, in the case of the highest organisms, it sustains by prodigal expenditure, during a period of helpless infancy and dependent adolescence.

Nor is Death to be denied aught of the grace and beauty of this descent and costly sacrifice, aught of the sweetness of expiration — the incense of its consuming flame, since these truly belong to our mystical thanatopsis. We close our eyes only to the weakness and decrepitude, to the rust and ashes, to the mere outward accidents that disguise the might and kindliness of Death.

* * *

The mystery of Evil is bound up with that of death, and the considerations already advanced respecting the one are alike applicable to the other. The mere body of Evil, like that of Death, is the after-part of a mystery far withdrawn from outward obser- vation into the unseen depths of creative The Mystery of Evil purpose, as the secret of winter is hidden, beneath its white frosts and behind its dun skies, at the very roots of things in the earth and in the heav- ens, and is not disclosed in the falling leaves or in the cold blast that sweeps through the naked forest. In our mystical vision Evil is seen to be essential to life —to its tropical movement of flight and return, hidden in its nascence and aspiration, and in its descent in- wardly beautiful and gracious, looking toward renas- cence; being in reality one with Death in its intimate association with the glory that is unseen, and with the pathos of all earthly experience, whatever may be its outward disguises and contradictions.

Even Sin, which is the sting of Death, must have its reconcilement with eternal life. We turn from the raggedness, the vileness, and the emaciation of the Prodigal, and regard only the unseen bond which brings him home, while we hear a voice saying : *This my son was dead and is alive again, he was lost and is found.*

Here, too, we but follow the mystical imagination of science, seeing in the spiritual world an attraction as mighty and as effective as that of gravitation in the physical; and, like Newton, we turn from the acci- dental appearance of falling to the unseen reality the mystical drawing to the heavenly centre; we turn from the weight that seems a burden to that which in the new interpretation becomes "an eternal weight of glory."

SECOND BOOK

NATIVE IMPRESSIONS

WHAT was the earliest thought of Death? The most primitive religious cult of which we have any record was the worship of ancestors. This carries us back to a time when in human thought there was no distinction between humanity and divinity. Man was a god in disguise, wearing the masque of Time, and Death was the unmasquing of his divinity. Evidently this ancient imagination was in no wise misled by the diminuendo of a descending movement that seemed to end in utter weakness; the vanishing point divided apparent impotence from an infinitude of power. To pass wholly into the unseen was to re-enter the latent ground of that potency of which the visible world was the manifestation in a continuous creation; and, in this restoration of higher power, there was no obliteration of personality but rather an enhancement of it, so that the pulsations of the universe seemed to be from stronger hearts than beat upon the earth. The mighty resurgence of life in dawns and spring-times was especially and most intimately associated with the dead — it was their Easter. Thus it happened that trees and indeed all plant life came to be thought of as mystically expressing the newness and elastic upspringing of life

[Marginal note: Native Impression of Death]

[Marginal note: The Dead Mightier than the Living]

that had been buried out of sight, buried like the seed
which dissolves for germination, sown in weakness
and raised in strength, sown in corruption and raised
in incorruption. The golden myrtle bough which Virgil
makes Æneas pluck before he can descend to Hades is
a survival of the old association, and primitive folk-
lore abounds in similar instances.

The serpent, because of its complete exuviation and
brilliant juvenescence every spring-time, was a characteristic symbol of underworld divinities, who presided
not only over the nascence of all things but over all
increase and fruitfulness. Even in the later mythology Pluto was the god of wealth.

The reader will immediately connect all this with
what has already been presented as the mystical vision
of Death, and see how accordant with that view was
man's earliest impression.

The modern habit, into whose texture enter so many
and so varied strains of sentiment, thought, and language, is closely wrapped about us, and is quickly
adopted by each new generation, so that we have quite
lost the native sense of things ; and even so much of
it as lingered about our infancy is irrecoverable by
us save in the faintest reminiscences. The scarcely
awakened sensibility of the child of to-day is forth-
with clad in raiment ready-made and thrust upon it,
and confronts elaborate artificial structures that confine it in many ways, while in others it is stimulated
by suggestions forcing it into the vast perspective of
intellectual and æsthetic symbolism. In rare instances
is the child saved from this too hasty investiture by
fortunate neglect or the still more happy circumstance

of solitude in the presence of Nature, and so enters into the kingdom of the naive ; and in all cases he has some protection through the long, slow waves of feeling that resist invasion and fraction. But generally these muniments of childhood's native realm are soon broken down, and such impressions as are won in their naked purity are rapidly dissipated.

It is difficult for us to abolish our perspective, and such impressionism as we have in recent art and literature is so remote from native sensibility that it belongs rather to the end than to the beginning of things, to the *fin de siècle* than to a primitive age.

Poe and Maeterlinck are far removed from Homer, who himself belongs to a period representing the youth of the world, not its infancy. The impression of death in Poe's poem *The Raven*, while it is more subtle than that given in Maeterlinck's *L'Intruse*, is not naïve — it is the reflex of experience. The native intimation is more truly conveyed in De Quincey's infantile association of his little sister's death with the crocuses than in "the silken, sad, uncertain rustling of the purple curtain" and all the other shuddering sensations inspired by Poe's bird of ill-omen. The refrain of *The Raven* is "Nevermore." But to the native sensibility Death is not an alien or an intruder ; nor are the Powers of Darkness unfriendly, being the true Eumenides, promising always bright returns. That which is taken from the light is hidden in the quickening matrix. The last gift of vanishing life is a seed, suggesting at once burial and germination. Thus the many-seeded pomegranate was the pledge between Persephone and Pluto. A sculptured slab recently excavated in Attica

shows the Eumenides in their most archaic representation, before they were transformed into Furies. They
are figured as benignant goddesses, each holding in
one hand a serpent and in the other a pomegranate,
and before them stand a young husband and wife, expecting a blessing.

The later pagan mythology was as wide a divergence
from primitive impressions as is dogmatic theology
from early Christian feeling. The rude infancy of
humanity left of itself no record, and there is little to
reward our most diligent quest of the naïve. The
savage races of to-day are degenerate, and their inveterate simplicity more completely veils the native
sense than does the complex environment of more aspiring peoples ; even their myths, handed down by
tradition, lack the naïveté of the Indo-European. The
retention of the native in indigenous races, in those
secluded from contact with others, and in those whose
development has been arrested, holds only the desic-
cated semblance, like an embalmed mummy ; and the
return of the native in degenerate races is no true restoration, belying and contradicting its original, being
indeed the more fallacious because of a fancied re-
semblance. The wildness of an old garden, once cul-
tivated but now come to decay, bears no true likeness
to the wildness of native flowers.

The archæological researches of this century have
given us some glimpses of a quasi-primitive humanity,
mere fugitive hints which, after all, are not more significant than those furnished by old Hebrew scripture
in certain passages caught and held there from some
otherwise long-forgotten past.

II

The childhood of a race has this in common with
the infancy of an individual—that its larger familiarity
is with the invisible ; it is naturally mystical. The primitive man has not that facile   Native
                                                   Mysticism.
handling of things which takes away their
wonder, nor that ease of thought and speech which
provides for him a fund of loose words and notions
which he can toss to and fro daringly and at random.
A look, a spoken word, an idea, a dream, is fatally real
to him, for good or for evil ; and he invests everything
about him with an ominous significance. Tokens have
not become common coin. His industry is concerned
with living things, with flocks and herds. In his commerce values are real, not merely representative. To
him Nature lives in every fibre of her being, nothing
is motionless or insensate ; it is a flowing world. No
masterful meddling or violence on his part disturbs
this impression. The growing tree is not to him something to be thought of simply for his use ; the forests
are as free from his invasion as the clouds above them,
and the streams pursue their course without diversion
or disturbance. There is nothing to break the living
veil of illusion—a shimmering veil of lights and shadows, of comings and goings, pulsing with the beating
heart of the Great Mother, whose changeful garment
forever hides and forever discloses the charm of her
wondrous beauty. In the free play of this sincere life,
where his naïveté answered to the perennial freshness
of the world, there was no room for the unreal play.

3

No sharply defined perspective furnished the ground
for distinction between small and great, high or low.
There could be no idolatry in the *Magnificat* of a worship that exalted the meanest creature. The sublime
superstition which lifted the lowest phenomenon to the
highest plane had nothing in common with what we
call superstition, whose omens are fortuitous and trivial, and whose signs have lost their significance. To-
temism (as we understand it), fetichism, witchcraft, and
sorcery are perfunctory relics of what was once a living
correspondence. We juggle with the dry twigs of what
was then the green tree of life. All that we imagine
as possible in clairvoyance was more than realised in
the primitive sensibility, not as yet disturbed and confused by those facile mental processes which loosen
the bond of the eternal familiarity.

When appropriation was limited to living uses, the
possession of things was not tenacious enough to imprison the soul in an artificial environment; and thus
inward meanings were conserved in their newness. In
this regard of the world the new was still the old, the
surprise deepening the sense of familiarity. Time itself,
in the childhood of the world, is the reflex of eternity.

When only living uses were regarded, the seizure of
man upon his earthly kingdom was eager, swift, and
passionate, but the reaction was quick; that which
was grasped was readily released. It is only against
the deep backward abyss that desire is a longing,
looking forward to untrodden ways, to a tale not yet
told, and yet falling back into the darkness as upon
the infinite source of its strength, with unfaltering faith
in resurgence.

III

It is peculiar, therefore, to primitive man that the backward look seems dominant, even in eager forward movement. Tenses are confused, as in the Hebrew the past is the prophetic tense, and as in our Anglo-Saxon the term *was* is the inten- sive form of the present, meaning *still is*, and so is caught *passing* into the *future*. That of the stream which has passed is that which has gone forward. In this primitive paradox and confusion (which is, indeed, characteristic of all real thinking) we have the feeling of a flowing world, whose end is its beginning, as the ultimate of a plant is its seed. The prominence given to memory and tradition in the early education of a race is not for the sake of stability, but is rather the regard of a growing tree to its roots, whither its juices perennially return; it is fidelity to the ground of quick transformation. This backward look is evident in the phrase used in patriarchal times, saying of a man when he died that he was "gathered unto his fathers." Therefore it is that among primitive peoples we find no allusion to a future state. The idea of recession, of return, dominated the native impression of all tropical movement. The blood was the life, and, wherever shed, it returned to its source, as the waters returned to their springs. This tidal stream or life current of humanity (limited in the primitive conception to the family, or the *gens*) found its way back to the well of its issue. Thus kinship was the first of all sacraments, the fountain of all

The Backward and Downward Look

obligation, so that all sin was a kind of blood-guiltiness.

To this natural piety was joined a natural humility.
The tree of life, while it grows upward and its unfolding leaves rejoice in the light, never loses its fidelity to
the darkness nor the habit of its descending juices.
The intimate association of man with the earth was
the largest reality in primitive faith, Semitic or Aryan.
The earth was the mother of all living, and the earliest
idea of divine as of human kinship was one deriving it
from motherhood rather than from fatherhood.   Solar
and astral worship belonged to a somewhat later development, when human thought entered upon a lárger
range, taking the stars into its counsels, as is indicated
in the term *consideration*.   Desire, in its earliest direction, was earthward, away from the stars—*desiderium*.
The sun first entered into the sacred drama through
his association with the earth, through a divine hus-
bandry corresponding to the human ; and in the darkness this association was continued through his participation with the Great Mother (Isis, Rhea, Cybele,
Ishtar, Demeter, or by whatever name she was known)
in the dominion of the underworld.   The sun-god was
ever a ministrant hero, like Heracles undergoing mighty
labors, and finally overborne by death, becoming a
theme for such passionate lament as wailed over autumn
fields in the song of Linus or the requiem of Adonis.
But in the Demetrian worship of primitive Attica
even this pathos was associated with Persephone, the
daughter of the Great Mother—so much nearer to the
heart of man, in these earliest mysteries, was the earth,
so much more impressive the sorrow of maternity!

From the Powers of Darkness and not from those of
Light was friendly aid solicited in the earliest human
worship. The Titans were brought into alliance with
man before he lifted his eyes in prayer to Apollo.
Divinity had its home in the earth, and its haunts in
the springs which quicken the ground. Death opened
not the gates of heaven ; and even at a later period,
when God was exalted, as the Most High, into the
heaven of heavens, the translation of mortals to His
presence was exceptional. Paradise, like Sheol, was
beneath the waters, and it was possible to look from
one into the other. In the most primitive period all
men alike passed to Sheol at death, the idea of Paradise, like that of Elysium, being a later conception,
when penalties and rewards, as the result of a divine
judgment, came to be associated with a future state.
Indeed, as we have seen, the domain to which death
introduced the soul was thought of as past rather than
future—the estate of the fathers.

It is not easy for us to even ideally reproduce a period when men lived in a primary field so directly vital
that their uprightness seemed to them like that of a
tree, a living righteousness, having no consequence
save in its fruit, the ultimate of which is expressed in
its seed ; when they looked upward by feeling downward, and forward by feeling backward ; when not only
the springs of life were divine, but its whole procedure
so entirely of divine ordinance that to think of it as a
probation or an experiment would have seemed blas-
phemous. The sense of a real Presence, holding them
by an inevitable bond, forbade conceptions quite germane to modern experience, when men think of them-

selves as the arbiters of their destiny. In the primitive thought good and evil, blessing and damnation, belonged to life, as such, from its beginning, even as light and darkness, pleasure and pain. To the native impression fear is as natural as hope, sensibility itself having its beginning in tremor and irritation.

This view of primitive man is quite as mystical as was the primitive man's view of life, and is largely the product of our imagination. We can only ideally reproduce absolute realism, and the men who had most absolutely the historic sense are themselves prehistoric. The native man is as much a mystery to us as a man born again seemed to Nicodemus. He is not the man we know, and the attributes we have been ascribing to him belong rather to dormant humanity than to a progressive order. What amazing stupefaction of abysmal slumber must have still held in suspense all the proper activities of manhood in a being who looked down to his God; who confounded the divine life with that of every living thing, looking indeed upon the lower animals, and even upon trees and stones, as somewhat nearer divinity than was himself; as if he must reverse the stages of his own antenatal evolution, in order that through the mediate series he might find the way to Him who was the Most Low !

IV

The earliest spiritual lore was from the education of sleep—of this very sleep which in the typical primitive era withheld man himself, as in every new generation

it withholds the infant, from merely outward meanings
and uses, and within the realm of a divine
mystery. What man was to be in his mas-   The Educa-
                                          tion of Sleep.
tery of the world was a destiny hidden from
himself—a destiny dominating him even while his ancient nurse and mother clung to him and often drew
him from the light which dazed his eyes back into her
helpful darkness. Indeed, it was from her bosom that
his strength was nourished for flight ; she was at once
Lethe and Levana, giving him sleep and also lifting him
into the light. The lusty outward venture would have
seemed too perilous but for her helping hand, and the
visible world alien and fearsome but for her whispered
names of new shapes, linking them with an older wisdom. His infancy was thus the period of divination.
Naturally, therefore, he thought of death as divinisa-
tion — not as an exaltation through some starward
movement, as the apotheosis of a Cæsar seemed to
the Roman, but as the restoration of latent powers
through descent and by way of darkness.

We who know only the Hades of later mythology,
peopled by bloodless shades, weak wanderers shiver-
ing between two worlds, being neither wholly alive nor
wholly dead, but held in the vain suspense of an empty
dream, forget that, in the earliest thought of men, the
dead were mightier than the living. The worship of
ancestors was the offspring of this impression. Men
covenanted with the dead as with the gods, and believed that they thus availed of the larger potency
and wisdom of the departed. The sword of an ancestor in the hand of his descendant had an access of this
superior energy.

In this time, when man especially leaned to the darkness, he found the way to unseen springs of power, ancestral and divine—a direct and sure way, familiar then but afterward forgotten or obscured. The spells of sorcery and necromancy were the perversion of this living ritual by which man once courted and won the Invisible.

All rituals grew out of this primitive ritual, known as the Way, but, losing the living reality, degenerated into meaningless routine. The profound meaning attached to the Way in all Oriental religions represents inadequately the original meaning. The plant knows the way to the water-springs. The habit of animal instinct, repeated from generation to generation, implies the divining of its way of correspondence. The ancient gathering of "simples" was the following of a path as sure and as mystically familiar as that which led to the means of nourishment. This Way began with the beginning of an organism, of an embodiment whereby the desire of the spirit became the desire of the flesh. The hunger which shaped the mouth informed it with a selective wisdom, whereby it found its response in a world it had always known, being outwardly stimulated and helped by a world which had always known it. The familiarity whereby Desire finds its Way in the visible world, blindly recognising, courting, and winning its respondents, which on their part are also seeking and finding it with the same blind insistence, is nourished in the darkness that is the background of all existence in time and in the world. Thus the Eternal Bridegroom is met, in all His myriad disguises, in the realm of His beautiful illusions; but in death, when one turns back into the darkness, all disguises are laid aside and He is seen face to face. And, as consubstantiality is the ground of correspondence in the visible world, Death is an awaking into His likeness.

Such was the native impression of Death. The evanition from the light into the darkness, recovering eternity, could not be for the primitive man the occasion of doubt or solicitude ; it was the ground of faith, through a covenant older than time.

Whenever any remarkable revelation was to be made to man he was brought into " a deep sleep." The ordinary occultation of the world in night and sleep became for him the supreme season — *suprema temp sus dici*, as it was phrased in the old Latin sacred books. Sleep was the undoing of all in man that grew in the daylight, and a committal of him to invisible powers which wrought in him their work, and from which there was an influx of divine wisdom :

> *In a dream, in a Vision of the Night,*
> *When deep sleep falleth upon men,*
> *In slumberings upon the bed ;*
> *Then he openeth the ears of men*
> *And sealeth their instruction,*
> *That he may withdraw man from his purpose,*
> *And hide pride from man*

In this occultation the sense of reality was enlarged rather than diminished, raised to a higher power, and a new world was created in a truer vision. The human was so intimately blended with the divine that the distinction between them was blurred, even as in death this distinction was completely lost. Accordingly the intimations of the dream were accepted as divine.

Wholly apart from the mystery of sleep and from the divine intimations of the dream, there was for man in this occultation the beginning of a spiritual philosophy. Sleep not only gave man a standing in a nearer divine presence, but the fact that life and thought went on when the body was motionless developed a consciousness of the human soul as independent of the visible world, and even of all that he ordinarily called himself. There was movement which was not locomotion, and a free play of mental activity involving an indefinite expansion of time. If there had been no night, a vague and fragmentary spiritual consciousness might have arisen from shadows and echoes. But in sleep the abstraction was complete, spontaneous, and inexplicable, and there was added to the independent existence of images their independent motion; there was a moving drama, wherein the self could become others, still remaining itself, being at the same time actor and spectator. There was vision with closed eyes, and hearing as with an inward ear; while the immobility of the bodily members seemed to be not merely the veil between two worlds, but the very condition of free psychical activity.

V

When the habit of abstraction, thus begun, became facile, the dream began to lose its importance as an especially real psychical operation; and its divine intimacy was loosened, until at length the easily shifting notion displaced the intense reality. A corresponding change affected the entire human re-

The Awaking.

gard of the world. Outward ends began to obscure inward meanings ; the primary became secondary ; the eternal familiarity yielded more and more to the temporal; that which had been the most intimate became alien. Man was fully awake, realising his peculiar destiny as a progressive conscious being. His philosophy, passing out of native impressionism, became, through notional abstraction, the ground of the exact sciences ; his language passed into its secondary meanings ; loose thinking came to be called close and rigid, as confined within definite limitations ; art, in like manner, passed from its purely vital field into that of representation, of images and similitudes ; the sacrament of kinship was weakened by the expansion of the family into wider communities ; and humanity flew out of its chrysalis, as a planet from its nebulous matrix. The dead and the divine became remote, no longer in immediate correspondence, but visiting men as ghosts or as angels — in either case still retaining their old divine designation as *Elohim*. The human cycle, distinct, self-conscious, and self-sufficient, sought completeness in the visible world, evading and denying the eternal. The conscious regard was mainly forward and upward, spurning the roots of the Tree of Life, looking rather to the fruit of the Tree of Knowledge. God had removed from His world to His heaven. Sheol was inhabited by weaklings, and death became in human thought the dread descent into that shadowy realm of impotence and insignificance.

The Heroic age as represented in Homer's Epics — especially in the Odyssey — had already lost the native sense of the invisible world and all homely familiarity

therewith. The Hades of the Odyssey is a world of
gloom into which the glories of the earth pass as into
a garden of faded flowers. When Odysseus, still belonging to the world of the living, is permitted to enter
the confines of this awful realm, a throng of pallid
spectres presses forward with insane hunger to drink the
blood of his propitiatory sacrifice. He sees Achilles,
and the burden of his old comrade's speech with him
is envy of the joys of life in the cheerful light of day.
The western sea bordering this underworld—the element of water itself being associated with dissolution—
was the haunt of Gorgons and Chimæras, of Circe and
the Sirens, whose charms and sorceries wiled men to
nameless degradation and ruin.   Homer's Poems and
the great Hindu Epic — the Mahabarata — show the
Aryan race at a much more advanced stage of civilisation than is generally supposed ; and one important
evidence of this is the fact that already the Powers of
Darkness have been submerged and are held in awful
abeyance.   The Eumenides have already been transformed into avenging Furies.

The Babylonian conception of the underworld was
even more degenerate from the primitive idea.   Our
first historic acquaintance with Phœnicia and Chaldea,
as with Egypt, is at a time when these countries are already famous for mighty cities, engaged in commerce
and in manifold industries ; and to their peoples the
thought of the world beneath the waters was like that of
a vast necropolis, whose dusty ways are untroubled as
in the suspense of an endless dream. Yet there was no
contrasting idea of heaven as a possible abode of mortals after death ; all alike must pass from the life of a

sunlit world to this realm of shadows. The earthly
aspirations of living men, in the full tide of youthful
strength engaging every energy in the accomplishment
of definite results, were jealous of invisible powers,
whose work seemed a negation of their own positive
constructions.

This apparent denial of Death was an illusion nourished by the very powers which it sought to thrust into
outer darkness and oblivion—nourished especially in
the heart and conscious thought of man, because it was
his peculiar destiny to express to the uttermost the
earthly mastery and the temporal familiarity; to lose
himself in the monuments of his art, whose duration in
time seemed a blazoned contradiction of eternity; and,
like one in a dream, to be buried in his terrestrial
economies.

The denial began with the first conscious progression—the first lapse from instinct into rational processes, but it was completed only when man became
wholly absorbed in his Time-dream, when, with eyes
closed to the invisible world, he came to think of that
world as itself dormant and oblivious. The Eternal
taking upon itself the masque of Time, so man, one
always with the Eternal, became a part of the masquerade, contributing to its delicious and painful bewilderment through disguises of his own, in the deepest sense inhabiting the world. And Death was the
master of the revels. In his secret heart is lodged
the power of a resurgent life, even as it is Lethe who
is the mother of Memory. He it is - this invisible
Angel of Life—who out of the rich darkness puts forth
the blade and bud and babe ; all the fresh and tender

luxuriance of growth is but the imagery of his abundance. His potence is the hidden spring of youth. But
also it is he who is confronted at every turn as a smiling wrestler inviting to conflict ; he who uplifts appearing to the outward vision as one who threatens a fall
—an archer inciting to protection against his own arrows, to wariness against his waiting destruction. To
man lost in the things of time, he who is the Deliverer
appears as Gaoler—he who alone faces The Real as
the King of Shadows!

VI

But to the primitive man—at least to our imaginary
type, never, indeed, in any record, known to us as
wholly free from the outward entanglement
—Death and the underworld were not held
as thus irreconcilably alien, nor as thus
shorn of their might.

Virtue or
Annihilation.

The native impression, on the visible side, regarded
the universe as a living reality—the diversification of
the divine life—and, on the invisible or vanishing side,
felt the elastic tension and expansion of that life as a
vaster reality. This impression was not confined to
the term of an individual existence begun at birth
and ending in death, but embraced all appearance
and disappearance, having a sense of constant pulsation, in which there is always a coming and going, as in an ever-changing garment that is being
woven by a shuttle now darting into the light and
then back into the darkness. This reflex movement, as connected with vanishing things — with all

things as momently vanishing — spontaneously re-
bounded to the central source, and was not interrupted
or distracted by any too fixed regard of the external
world, but rather took that world with it on its refluent
tide, bathing it forever anew in the pristine font of an
eternal life.

In the dissolving view disappearance was not merely
negative; it was more positive than appearance. It
was from the ground that Abel's blood cried unto
the Lord. Something of this feeling remains among
the Chinese, who having written their prayers upon
paper, then burn the paper, having more faith in the
obliteration than in the literal expression. There is
marvellous virtue in annihilation. The mystery of the
universe can be nakedly disclosed only in the death of
the universe; nevertheless it is the mystery of every
moment of every living thing—lost in the life of that
moment and recovered in its death.

VII

We dwell upon this native sense of the wonder
which life has in its fresh and radiant appearances and
its more marvellous vanishings, because it
helps us to see how natural is that transcen-   Native Sense
                                                 of Reality
dental mysticism which by elastic rebound
overleaps the apparent finality of death; which finds
in the point of rest the initiation of a miraculous motion, so that zero becomes the symbol of the Infinite;
which has such faith in Life as to give no credence to
its apparent diminutions as signs of weakness, seeing

in them rather the intimations of some mighty transformation already begun.   Such a miracle was witnessed in an eclipse of the sun—especially in a total
eclipse, when complete annihilation seemed to be followed by renascence.

It is very difficult for us to even imagine this native
mystical apprehension of an eternal life. We have the
impression in some degree awakened in us by vast barren places, by the immobility of landlocked waters, by
the silence of deep forests, and in seasons of unbroken
solitude. It is not a sense of lifelessness in these situ-
ations, but of deeper life suggested through the absence of color and sound and motion, which are usually
so prominent in our perspective. In the outward silence
the inward Voice is heard.   To us, perhaps, the Voice
seems alien, but to the primitive man it was that of a
Familiar. We shrink from intimations which he courted, his solicitation having become for us a dread solicitude ; and the Way frequented by him—kept open between him and his ancestral home—we seek to close,
setting a seal upon every sepulchre, barring out the
revenant.   In spiritualism and occultism we attempt
an awkward coquetry with vanished souls—and in this
casual necromancy how antique, indeed, seem our cor-
respondents, even the nearest of them ! In insanity
there appears to be an abnormal restoration of the
atavistic channel. How significant, then, it is to note
that there was a time when, in a sane mood and without jugglery of any sort, the living had communion with
kindred souls departed — a cherished intimacy which
made the darkness friendly and as fragrant as the
breath of love, and which with resistless charm drew

them within the shelter of overshadowing wings, with
in the circle of fatherly and motherly might and
bounty.

VIII

The naturalness of this mysticism distinguishes it
from mediæval and modern mysticism. In the primitive view, while the unseen was the larger
reality, the visible world was not less real, *Mediæval and Modern Mysticism.*
nor was the fresh and eager desire for that
world in any way suppressed or deprecated. Its sublime negation, whereby that which passed from vision
entered into a new and greater glory, had no likeness to the Buddhistic Nirwana, though it may have
been identical with the earliest meaning of Nirwana as
entertained by the primitive Aryan. Modern religious
mysticism is not content with the natural transcendency
of a transforming life, and is therefore disposed to sacrifice Nature to the supernatural, so that its consideration of the external order of things, whether as divinely or humanly ordained, falls into the slough of
pessimism. Only the blood that leaps into the quick
and full pulsation of earthly life can have an elastic rebound to its eternal font. The sense of fatherhood
and motherhood, imperatively linked with the sacrament of kinship among all primitive peoples, could not
have tolerated the Tolstoian view of marriage. Only
artificial uses were excluded from primitive life, and
even these lay ahead of it as inevitable in the natural
course of progress ; but, these not yet existing, the
abuses of convention prompted no revolt like that
which enters into modern speculation.

4

The denunciation of selfhood which is the key-note
of all modern mysticism could have had no place in a
primitive estate, in which selfishness had no expression
save as the natural postulancy of childhood—a great
hunger to which all things responded. The need most
real was that of fellowship. Exiled from his fellows,
man in the presence of Nature experiences a strange
sensation. We say that a man is born alone and that
he dies alone; but he is born of his kind and to his
kind he dies, so that, in either case, fellowship is emphasised. But, in human embodiment, confronting
the physical world, unsustained by human companionship, his loneliness is supremely awful, and, if prolonged, would in time deprive him of reason and
speech and of every distinctively human characteristic.
Nature, to the solitary individual man, is dumb and her
ministration meaningless. In this situation he is mor-
ally and spiritually a nonentity; he can have neither
selfhood nor communion. He is not a normal animal,
but defective, degenerate man. The isolated man is a
man wholly, uselessly, irretrievably lost. Neither life
nor death has for him any meaning, and to him God
can in no way be revealed. He is nourished to no
purpose, increased for no proper function, and even his
diminution and disappearance seem anomalous. If
we could suppose him to have never had human fellowship, he would be even physically incomplete, a lost
half of a being, the dominant system of his cellular organism—an *imperium in imperio*—having no response
and mocking his empty arms, however much of the
world they might hold, despising his pain and travail
as utter vanity. Life would have no romance of its

adventure and the universe no prize in its treasure-house worth the winning or for whose loss one might grieve. Only he who loves can weep, and man loves not the world nor self until he has loved his kind.

Not selfishness, then, but sympathy is man's native feeling. Only in a fellowship can he find himself, only in a human kinship the divine. The cosmic preparation, outside of himself and in his own organism, is not for an individual but for humanity; it is the foundation of loving fellowship and broad enough for universal brotherhood : indeed, the operations of the physical world as related to man can neither have their full effect nor be fully understood save in such a brotherhood. The preparation is for love. The very diversity of individuation, the apparently sealed envelope of separate embodiment, forbidding fusion, stimulate association and enhance its charm. The first man-child born into this fellowship may become his brother's murderer ; ambition may produce dissension and promote violence, and the very closeness of family and tribal relationship may lead to conflict with other equally solid leagues, and so appear dissociative : but, in the end, crime, oppression, and war will compel larger solidarity and ampler freedom. The enlargement may substitute conventional for natural bonds, but within the scope of the widest convention there will remain the family on a surer basis, and the social activities in their freest sympathetic expansion ; and thus Love that seemed to be hidden will remain lord of human hearts.

In any period, therefore, of human progress, selfhood is but the reflex of fellowship, first human and then

divine, or rather both in one.   A subjective mysticism,
contemplating as possible the exclusion of selfhood by
an influx of divine life, is irrational.   It is the expansion of selfhood, the deepening of its capacity through
its exhaustive demand upon all ministrants, human
and divine, that at the same time provides a guest-
chamber for the Lord and an abundant treasure-house
to be exhausted in ruinous expenditure for the service
of man — a service most effective when it most truly
expresses selfhood.

Since all religious mystics, of whatever creed and of
whatever race, have, from the beginning of a philosophic
era, agreed in this assault upon selfhood, their unani-
mous expression commands respect.   The general as-
sent to a proposition, as, for example, that the sun re-
volves about the earth, does not prove the truth of the
proposition, in the absolute sense, but it does indicate
a general impression as its real and true basis.   What
impression, then, is it that has been so generally entertained as to be the real basis of this mystical revulsion
from selfhood?   The word *mysticism* is from the Greek
muesis, the closing of the eyes—that is, one turns from the
sensible appearance, shuts his eyes to the visible world,
in order to see true.   Some fallacy, therefore, some inevitable delusion, is conveyed to the soul through the
appearances of things to the eye of sense, something
which must be corrected, even reversed, in the spiritual
vision.   The spiritual is thus opposed to the natural,
even as the Creator has a perfection as opposed to the
imperfection of the creature.   The universe stands in
contradiction to its source—the natural manifestation
opposed to the spiritual principle.   How readily has

this radical distinction between the creature and the
Creator commended itself to the prophet and spiritual
philosopher of all ages ! " Yea, the stars are not pure
in His sight. How much less man who is a worm !"
" There is none good but one." If a man turns from
the entire visible world to such truth as can be only
spiritually discerned, shall he not also turn from himself, making the vastation complete? If nature is an
ignis fatuus, misleading him, how much more deceptive
the imaginations of his own heart !

There is in this impression the deepest of all truth
both as to insight and as to action : as to insight, because it is the comprehension of evil as associated with
all manifestation, divine and human ; as to action, because it is a recognition of the necessity of repentance
and regeneration to all the transformations that have
ever been or ever shall be wrought in man or in the
world, so that the universe itself is forever being repented of and created anew—the new creation being a
redemption.

The truth thus stated brings the impression resting
upon it into accord with the native and natural mysticism ; the evasions and perverted expressions of it
have reflected the errors of existing systems—such errors as were illustrated in Oriental dualisms (most no-
tably in Manicheism), in Neo-Platonic speculations,
like those of Philo Judæus concerning creation as
the work of good and evil angels, and in much of mediæval and modern Christian theology. All these errors illustrate the fact that philosophy, even as a part
of theology, is in its development not exempt from the
evil which is inextricably involved in all manifestation,

and so is something to be itself forever repented of and born again; these errors being a contradiction of the spiritual principle from which they are the departures.

The central principle of all systems, divine or human, impels the departure and demands the return, thus involving the destruction of every edifice that is builded; it gives into the light and takes into the darkness; it determines the maturing strength, the tenacity of structures, the consistency of systems, and it determines the dissolution also of all embodiments, for renewal and transformation. He who would forever hold to the structure, losing himself therein, and looking not to the source of life, is in prison, and for him the illusions of the light become delusions; but, on the other hand, he who turns from his dwelling save for new and brighter dwelling, who seeks the darkness save for the renewal of desire, who in expectation of immortality denies resurrection as fresh embodiment, sets his face against the mortal hope, and for him there is only the prospect of some level world in which there is no world to come. But Life knows no such sterile issue, and into whatsoever chamber the Bridegroom shall enter, again he shall go forth therefrom, rejoicing as the strong man to run a race!

The ultimate mysticism will be that of science vitalised by the Christian faith and of that faith illuminated in all its outward range by science; and it will be seen to be one with native intuition, but including a perspective commensurate with the visible universe. Christianity will again accept Nature, as indeed it did in its prime, holding it to be one with the Lord, and find in

its wonders as disclosed by science the counterpart of
the glory revealed in him ; while science, which is already insisting upon so much that no man has ever
seen, will translate its invisible elements into the living
language of faith.

The sequestration of spiritual life as something by
itself, apart from the life of the world and incommunicable therewith, is an exaltation that cannot be long
maintained, since the power of an eternal life must always be manifest in the freshness of time, in the renewal of the world. A new creation is only a new
nature, having its own trope, its proper action and reaction, and the inseparable companionship of life and
death.

What new embodiment awaits us at death — that
death in which we have no part and that has no part
in us—we know not, but we know that it is only transformation. "We shall all be changed." A new
sensibility would, in this present life, reveal to us a
new universe. When we come to consider that what
we now know as sex and what we know as death are,
in the present order, only specialisations occurring at
their due time in organic development, we may comprehend a possible order in which these would have
no such meaning for us—some such order as our Lord
intimated when he said of the children of the resurrection that "they shall not marry nor be given in marriage ; neither shall they die any more." But change
itself, unspecialised Death, these belong to any life,
as does also the unspecialised essential ground of
what, in all manifestation, we call evil.

IX

These considerations lead us to dwell more at length upon the native impression which regarded life and death as universal and inseparable.

The primitive man made no distinction between the specialised and the unspecialised. The vast background of the unseen to which he was conjoined by ancestral familiarity, and which therefore had for him only homely and friendly aspects, was very near, an intimate council-chamber to which he still had ready access and from participation in whose eternal decrees he had never been excluded. Here it was that Love and Death and Grief had been assigned their part and place in the cosmic harmony.

Universality of Death.

In the visible foreground—to the primitive man a very narrow field, in which a mere fragment of humanity confronted the mere fragment of a world—were to be enacted the mysteries of the ancient council-chamber, represented in masquerade, wherein the old meanings were to some extent disguised, but by a veil far more transparent than that which we have clothed them with in modern thought and custom. Between the visible and the invisible there was a frank and easy interchange, with no strain of religious awe, no logical embarrassment, no grave solicitude. The human, the natural, and the divine were blended into one very simple drama, from which we would turn in mental, æsthetic, and moral contempt.

There was no distinction such as we make between living and non-living matter. The whole universe was living and sentient; and so persistent was this native impression of an animate world that it was entertained for centuries by philosophers, and even by Kepler, who first formulated the laws of planetary motion. The domain of death was coextensive with that of life: Nature was not only living in every part, but in every part also dying. In this earliest faith even the gods were mortal. That sacrament of kinship in which love and death and grief were first known to the heart of man, and known as inseparable, was a covenant which had no limitations. Divine love, like the human, was without death unavailing, lacking its crowning grace.

The Olympian dynasty of gods, hopelessly immortal, was a later conception; and this dynasty represented relentless law and force, loving not man, nor coming within the pale of human sympathy. During the whole period of ancient paganism, the human heart turned from these passionless divinities to those of their sacred mysteries—to gods who could die and grieve.

The first estate of paganism extended the intimacy of human kinship till it included the visible universe. The fire upon the hearth-stone was but a spark of the flame of Love that spent itself for all needs. The bread and wine that gave strength to man were symbols of the largest ministration—a descent and death for human increase. The mother, who brought forth children from her body and from the same body nourished them, was the type of the divine motherhood, whose bounty was freely exhausted for all, even unto self-desolation.

In such a faith there could be no rebellious complaint against pain and frailty and death; the ab

sence of these would have confounded men, making
of life a nondescript, a shadowless, glaring absurdity.
Nearness to life, in this native feeling of its reality and
universal pathos, brought a reconcilement of its contradictions, and the exclusion of any element would
have disturbed its harmony, even though that element,
seen by itself, might have appeared discordant.

The primitive faith accepted death and evil, as it
accepted darkness and frost, and at the same time regarded them as parts of Love's cycle. Thus it emphasised the limitless divine bounty and indulgence, and
had no conception of human or divine justice. Pain
was not penalty. Blood that was shed called for
blood, but, outside of the bond of kinship, the voice
was silent, alien, untranslatable.

X

The social order has progressed through stages involving a constant and ever-widening departure from its
first estate of comparative simplicity and natural piety.

While man is pre-eminently a social being, the first
and natural bond of flesh and blood kinship is so intense,
reinforced by its vitality confined within a narrow field,

The
Weakness of
Primitive
Paganism.

as to seem exclusive and dissociative. The
parent has a jealous love of offspring which
makes even a neighbor seem alien and hostile. How much stronger must be the native feeling of a community thus bound together toward others not included in this alliance! There is in
this feeling a strange mingling of fear and curiosity.

The desire for communication will in the end overcome the jealousy. The most interesting feature of the earliest historical records recently brought to light by archæological exploration is the frequency of messages exchanged between princes of peoples widely separated, indicating also exchanges of visits and gifts and often intermarriage. Travelling was an ancient passion, and the eagerness with which the Greeks at their Olympic games listened to the foreign gossip of Herodotus has been characteristic of men in all times. There is in the satisfaction of this curiosity not merely the charm of novelty, but an indication of that amicability which is the ground of hospitality. In the beginnings of commerce a certain shyness was apparent—as in the custom of leaving articles of barter at places agreed upon ; and the fact that no advantage was taken of this shows how strong, in the crudest conventions, was the sentiment of honor between parties too timid to face each other in a mercantile transaction. Thus from the first there was indicated the germinal principle of a social order, based upon honor and justice, which was to extend over the habitable globe.

As the living bond was relaxed, surrendering its natural force for the gain of structural strength, the native intuitions belonging thereto were in a corresponding measure dissipated. The bond of kinship was physiological and instinctive, giving free play to the animal nature in the full range of its sympathies and also of its animosities ; but it is instinct that is submerged by rational and conventional systems, and hidden beneath the more complex operations that are its specialisation. The expansion was inevitable, resulting in the establish-

ment of a government quite different from the patriarchal, of treaties between peoples, and of internal police
regulation ; national consolidation ; empire as the issue
of conquest ; institutional stability, and the consequent
development of science, art, and industry : an organised
moral world.

Knowing how severe a strain primitive Christianity
has sustained in the material and intellectual development of western nations, we can readily understand
what havoc ancient civilisation made of primitive paganism. Among the Indo-European and Semitic peoples, the worship of ancestors was a dying cult in the
very dawn of that civilisation. The same intellectual
culture which banished the gracious ancestral divinities
brought in a dynasty which ruled the world by inflexible
law, and which was in accord with the social solidarity
based upon justice. The Sacred Mysteries were re-
tained, and with them the popular faith in a dying Lord
who rose again, and in a sorrowing mother, as also in a
sentient universe, which was inseparably associated with
the divine death and sorrow and triumph—so that there
still remained for the human heart a field of divine love
and pathos into which were lifted its own love and frailty, its passion and pain. But there had been a remarkable change wrought in this faith. For, while only in
the minds of a few had the ancient philosophy succeed-
ed in interposing an insensate mechanism between man
and God—a realm of matter, lifeless and deathless and
so cut off by icy barriers from human sympathy,—while
the scientific view which thrust the human heart back
upon itself, isolating its hopes and fears from their connection with the general course of nature, was not wide-

ly accepted by the people, owing to the limited diffu-
sion of knowledge, yet in the very development of a
complex order there was an inevitable tendency toward
this fatal schism ; and the idea of a future state as one
of rewards and punishments was generally adopted.
The recognition of a moral order under divine sanction ; the conception of retributive justice operating in
the future as in the present life, only with greater efficiency ; the distinct separation in the minds of men between good and evil, so steadfastly maintained that the
moral ideal implied the possibility of absolute rectitude
as the result of conscious determination, a perfectness
unknown to Nature and wholly excluding evil—these
were the results and reflexes of a social economy far advanced beyond its primitive estate and brought within
rational control ; and these modifications of the religious view served incidentally to reinforce the restraints,
however arbitrary and conventional, of civil government
and social custom.

Because paganism, in its earliest estate, was not based
upon the spiritual principle of universal brotherhood ;
because it never transcended the limitations of an imagination strictly confined to natural cycles forever returning into themselves, even as associated with the
unseen world, it was therefore irreparably damaged by
the incursions of a hostile philosophy, which preyed
upon its vitals, as did Jove's eagle upon those of the
Titan Prometheus. The destruction or the devitalisation of its material embodiment left it no place of ref
uge, since only in that embodiment had it a habitation.
Its disintegration could not be followed by rehabilitation from any principle within itself. As its action

in faith lacked the complete expression of a spiritual fellowship, so its reaction and contradiction in the outward social order was incomplete in the realisation of equity.

The structure of paganism, considered as a whole, in its religion and its outward economy, was, like its architecture, low-arched, too limited in its scope to escape ruin as a whole. It lacked the Master Mason to build it high, availing of weight for support, of descending movements for new ascents, of death for life. It was overweighted, and crumbled to the ground all along the lines of its construction, beautiful in its ruins, which in every part indicated a magnificent virile effort, and at the same time a fatal inherent weakness.

We shall see hereafter, when we come to consider the structure of Christendom, that whatever may be the departures of the latter from its spiritual principle—departures repeating and often exaggerating the defects of paganism—yet its scope is large enough for the completion of its cycle, through the consummation of its social and intellectual development, in a return to that principle ; and we shall also see how science itself in its later revelations helps to bring the human reason back to the recognition of evil—or what we call evil— as a reaction proper to life in all its manifestations, divine or human. The fraternal sympathy, which is the ultimate fruit of Christian faith, will restore, in new and higher meanings and appreciations, the universal pathology naïvely implied in primitive intuition.

THIRD BOOK

PRODIGAL SONS: A COSMIC PARABLE

THE DIVIDED LIVING

I

FORMLESS, imageless, nowhere, nowhile, non-exist-
ent—a Void; and over against this, all that is,
that ever was, and ever shall be—a Universe. Everything from nothing. We have no other
phrase for the mystery of Creation, save as
we express it personally in the words Father
and Son. For that which, in this contradiction between the essential and the manifest, we call Nothing,
for want of a nominative, is the infinite source of all
life. When we say of the visible world that it is the
expression of Him, we are saying as best we can that
the world is because He is; but even this idea of
causation falls short of the mystery, of which, indeed,
we can have no idea, since our imagination cannot
transcend the world of images. How can there be an
image of the imageless? We proceed through a series
of negations, abolishing time and the world, existence
itself, and when our annihilation is complete, the Void,
in our spiritual apprehension, brings us face to face
with the Father of Beginnings; the boundless emptiness becomes the boundless *pleroma*, or fulness.

Therefore it is that Death, which brings to naught,

5

discloses the creative power of life. If this power
were simply creative and not re-creative, formative but
not transforming, the world would be the seamless,
never-changing garment of God. From the first, in
all this cosmic weaving, Death is at the shuttle, completing the trope in every movement, every fold ; with
his face turned always to the Father, he whispers release to every living thing ; and thus he becomes the
Leader of Souls, bidding them turn from the world
that is, that he may show them a new heaven and a
new earth, calling them to repentance and a new birth.
He is the strong Israfil, winged for flight, and ever
folding his wings for new flight. Under his touch all
things turn—to noon and then to night ; to maturity
and then to age ; but we shall not find him in the old
which we call dead—*that* he has already left behind,
bidding us come and follow him, while with one hand
he points to a new generation upon the earth, and
with the other to an unseen regeneration.

Thus inseparably associated with the genetic, Death
is bound up with the mystery of Creation itself. The
evening and the morning were the first day.

II

Who can bridge the chasm between the unseen sub-
stantive in the grammar of Life and its genitive case ?
Who shall find for us the dominant in the musical gamut
—that original trope of genesis, through which the singing stars danced into the field of Dawn ? Who shall
show us the invisible fulcrum of the first leverage, the

initial of the celestial mechanics? There is no ship
we can make to launch upon the ocean which separates
the finite from the infinite, time from eternity, the world
from God.

There is, indeed, no such ocean, no such separation
—no chasm to be bridged. The web of existence may
have interstices; in time and space there are intervals
between things, degrees, similitudes, diversi-
ties; media that at once separate and unite. Creation· The
First Kinship
Here nearness and distance are comparative;
but no individual existence is near any other with that
intimacy which each has with the Spirit of Life; there
is no familiarity in the world like the eternal familiarity.
It is spiritually represented in the nearness of the eternally begotten Son to the Father; the Son is forever
Sent, yet is always in the bosom of the Father. The universe, expressed in the term Nature, reflects this intimacy; it is forever being born, flying from its source,
yet there is in the consistency of all its parts in one
harmonious whole no bond so strong as that holding
it to the Father. Procreation is the nearest image of
creation, involving at once otherness and likeness.

Existence seems a denial of Being, because we are
unable to predicate anything of Being save by the ne
gation of our predicates concerning existence. Moreover the progressive specialisation of existence seems to
involve successively more and more a surrender of the
potency and wisdom that, in the essential source of
all, are infinite. It is as if, in time and in the world,
the Father had divided unto all His living, every added
complexity signifying greater multiplicity and so a
greater division. The denial is apparent only. In

reality all visible existence is to invisible Being as the
stream to its fountain, so consubstantial therewith that
it should be thought of as one with rather than as related thereto, than related even as effect to cause. The
embodiment is proper to the spirit. The ever repeated
creation is genesis, a constant Becoming. The Eternal
becomes the temporal. The boundless life is the
abounding, and its bounds, or limitations, while on the
visible side contradicting boundlessness, are really the
bonds of kinship with the Eternal. The quality of
life is the same in the limitations as in the boundlessness. Finitude is of the Infinite ; Form is of the unseen shaping power ; and Transformation is essentially
genetic, creative.

III

In an unchanging world—if such a world were conceivable—we would have no apprehension of this genetic
quality of life, which is not suggested in a persistent appearance, but only in disappearance, or disappearance
followed by reappearance. That trope of a
cycle through which existence vanishes is,
therefore, a dissolving view fraught with spiritual suggestion. The end is lost in beginning. All transitions, all the phenomena of change,
become luminous points in consciousness, leading from
the fixed to the flowing, from ends to beginnings, from
the visible shapes passing before us to the invisible
shaping power ; and when anything so passes as to utterly escape vision — like the passing of a soul — we
have the deeper suggestion, from which arises a tran-

The Dissolv-
ing View : Its
Spiritual Sug-
gestion.

scendent mystical vision; a power is released in us
which follows the power that has been released, into its
unseen realm; and so we are ever pursuing that which
flies, even through the gate of its Nothingness, to apprehend, though we may not define, its essential quality, as our eyes follow the ascending mists till they vanish and we see the clear heaven, from which they are
no longer distinct, being one therewith and participant of
its powers.

IV

As through the trope which is Death is the entrance
to greater potency, so in that of Birth there is an apparent surrender of power, a veiling thereof in embodiment; and the first Genesis, if there were
a first, was the primary abnegation, wherein The Involve-
the Infinite became the Finite.   ment.

Standing at the gate of Birth, it would seem as if it
were the vital destination of all things to fly from their
source, as if it were the dominant desire of life to enter
into limitations. We might mentally represent to our
selves an essence simple and indivisible that denies
itself in diversified manifold existence. To us this side
the veil, nay immeshed in innumerable veils that hide
from us the Father's face, this insistence appears to
have the stress of urgency, as if the effort of all being.
its unceasing travail, were like the beating of the infi
nite ocean upon the shores of Time, and as if, within the
continent of Time, all existence were forever knocking
at new gates, seeking, through some as yet untried path
of progression, greater complexity, a deeper involve-

ment. All the children seem to be beseeching the Father to divide unto them His living, none willingly abiding in that Father's house. But in reality their will is His will—they fly and they are driven, like fledglings from the mother nest.

V

The story of a solar system, or of any synthesis in time, repeats the parable of the Prodigal Son, in its essential features. It is a cosmic parable.

The planet is a wanderer (*planes*) and the individual planetary destiny can be accomplished only through flight from its source. After all its prodigality it shall sicken and return.

Attributing to the Earth, thus apparently separated from the Sun, some macrocosmic sentience, what must have been her wondering dream, finding herself at once thrust away and securely held, poised between her flight and her bond, and so swinging into a regular orbit about the Sun, while at the same time, in her rotation, turning to him and away from him—into the light and into the darkness—forever denying and confessing her lord! Her emotion must have been one of delight, however mingled with a feeling of timorous awe, since her desire could not have been other than one with her destination. Despite the distance and the growing coolness, she could feel the kinship still ; her pulse, though modulated, was still in rhythm with that of the solar heart, and in her bosom were hidden consubstantial fires. But it was the sense of otherness, of her own distinct individuation,

The Prodigal Planet.

that was mainly being nourished, this sense, moreover, being proper to her destiny; therefore the signs of her likeness to the Sun were more and more being buried from her view; her fires were veiled by a hardening crust, and her opaqueness stood out against his light. She had no regret for all she was surrendering, thinking only of her gain, of being clothed upon with a garment showing ever some new fold of surprising beauty and wonder. If she had remained in the Father's house—like the elder brother in the Parable—then would all that He had have been hers, in nebulous simplicity. But now, holding her revels apart, she seems to sing her own song, and to dream her own beautiful dream, wandering, with a motion wholly her own, among the gardens of cosmic order and loveliness. She glories in her many veils, which, though they hide from her both her source and her very self, are the media through which the invisible light is broken into multiform illusions that enrich her dream. She beholds the Sun as a far-off insphered being existing for her, her ministrant bridegroom; and when her face is turned away from him into the night, she beholds innumerable suns, a myriad of archangels, all witnesses of some infinitely remote and central flame—the Spirit of all life. Yet, in the midst of these visible images, she is absorbed in her individual dream, wherein she appears to herself to be the mother of all living. It is proper to her destiny that she should be thus enwrapped in her own distinct action and passion and refer to herself the appearances of a universe. While all that is not she is what she really is—necessary, that is, to her full definition—she, on the other hand, from herself interprets all else. This is the inevitable terrestrial idealism, peculiar to every individuation in time — the individual thus balancing the universe.

VI

In reality, the Earth has never left the Sun; apart from him she has no life, any more than has the branch severed from the vine. More truly it may be said that the Sun has never left the Earth.

The Illusion of Distance.

No prodigal can really leave the Father's house, any more than he can leave himself; coming to himself, he feels the Father's arms about him—they have always been there—he is newly apparelled, and wears the signet ring of native prestige ; he hears the sound of familiar music and dancing, and it may be that the young and beautiful forms mingling with him in this festival are the riotous youths and maidens of his far-country revels, also come to themselves and home, of whom also the Father saith : These were dead and are alive again, they were lost and are found.   The starvation and sense of exile had been parts of a troubled dream—a dream which had also had its ecstasy but had come into a consuming fever, with delirious imaginings of fresh fountains, of shapes drawn from the memory of childhood, and of the cool touch of kindred hands upon the brow.   So near is exile to home, misery to divine commiseration—so near are pain and death, desolation and divestiture, to " a new creature " and to the kinship involved in all creation and re-creation.

Distance in the cosmic order is a standing - apart,

which is only another expression of the expansion and abundance of creative life; but at every remove its reflex is nearness, a bond of attraction, insphering and curving, making orb and orbit. While in space this attraction is diminished—being inversely as the square of the distance—and so there is maintained and emphasised the appearance of suspension and isolation, yet in time it gains preponderance, contracting sphere and orbit, aging planets and suns, and accumulating destruction, which at the point of annihilation becomes a new creation. This Grand Cycle, which is but a pulsation or breath of the eternal life, illustrates a truth which is repeated in its least, and most minutely divided, moment—that birth lies next to death, as water crystallises at the freezing point, and the plant blossoms at points most remote from the source of nutrition.

VII

We need to carry this idea of Death, as associated with Creation and Transformation, into our study of visible existence; otherwise the claims of philosophy as well as of faith are likely to be sacrificed The to those of a science which, in its persistent Tendency to Ignore the specialisation, tends to wholly ignore the Creative principle of creative life. We have no fear Principle of honest agnosticism, of dilettanteism, or even of infidelity. The real danger lies in the inflexible certitude of the specialist. The peril touches not religion alone, nor is natural science its only source. The extreme specialisation of modern life in every field confines thought as it does effort and tends to con civa

tion and stability.   Its perversity is in its opposition to
reaction ; it will not readily admit a solvent, and resists
every subversive or destructive element, unwilling to let
the dead bury its dead.   This tendency affects theology
more than it does physical, political, and economic science.   The children of this world are wiser in their
generation than the children of light, because they are
not so closely bound by unvital traditions, and also because a merely utilitarian interest compels solvency,
change, revolution.

The perversion of human thought, in its attitude toward Death and Evil, and its consequent exclusion and
ignorance of divine absolution as a constant and intimate creative transformation in Nature and humanity, is
especially easy to the modern mind which regards Nature
as impersonal and man's relation thereto as accidental
and temporary and mainly significant in its utilitarian
aspects.

Generally the terms of science are unvital.   Force,
matter, motions, vibrations, laws : these terms give us
no impression of a living world.   Science is confined
to a formal conception of existence, and is concerned
with quantity (the measure and proportion of elements
and their relations in time and space, mathematically
expressed) rather than with quality.   Even the theo-
logian thinks of eternity as duration, as quantitative
rather than qualitative.   The Latin for reason is *ratio ;*
and to the Greek all learning was *mathesis*, from which
the term mathematics is derived.   Next to the stress
which science lays upon the form is that which it gives
to uniformity, from which it makes those generalisations
that are called laws.   These limitations of science to

consideration of method and proportion are inevitable ; but since form is of the essence and quantitative relations have a qualitative ground, the true philosopher apprehends a reality beneath as well as in the form, the shaping power and wisdom transcending as well as immanent in the visible shapes of the world, and thus in every fresh scientific discovery he finds a new intimation of spiritual truth. All the manners of the universe become to him traits of the divine Personality in whom it " lives and moves and has its being." Too often it happens that the scientific specialist, when he transcends his specialty and enters upon the larger field of philosophy, brings with him into that field the unvital terms which are there inadequate and misleading. How, for example, can one who insists upon everlasting uniformity, and so upon invariable laws, express truly the spiritual apprehension of Life as a transforming power? The incompatibility is more conspicuous if these laws are regarded as impersonal, as belonging to matter, whether independently or by divine delegation once and for all, and, however imposed, as limiting the divine operation.

VIII

But all human specialisation, whether in science or elsewhere, follows Nature's own leading. We deprecate materialism, mechanism, and utilitarianism, but these are most conspicuous in the cosmic order. Man's development of outward structure, social, political, and industrial, corresponds to the cosmic development which prepared the way for

The Divine
Pattern of
Materialism

his progress, which, indeed, by the constitution of firmaments gave him a standing-place in the world. God is
the first materialist. Mechanism is celestial before it is
earthly and human.

Seeing, then, a world prepared for him, a world of
things ready for his arbitrary fashioning — metal and
stone and wood—things cut off from their living currents by natural sequestration, or which he might himself so cut off for food, raiment, and shelter, and, later,
for these uses in more ambitious and luxurious fashion ;
seeing, in his further progress, that he might lay hold
upon the living currents themselves and divert them to
his use in more complex and heavier undertakings, dividing them according to his requisition, or even holding
them in storage for his convenient and leisurely division ;
taking note, moreover, of a constant providence, answering to his prudence, and the regularity of Nature's habits, suiting a never-failing ministration to his needs—is
it strange that man should have yielded to the divine
temptation, conforming to the divine exemplar, the pattern shown to him not only upon every mount, but in
every depth and in every path opened to his eager
feet?

For, on the human side, there was not merely passive
yielding and conformity; there was desire, which seized
with violence upon a kingdom at hand. Save unto desire there is no temptation, no stimulation save of a
faculty, no ministration but to a craving capacity. All
embodiment is but the extension of importunate desire.
Man's entreating of the world is first and always a passionate entreaty; he " has no language but a cry." As
his embodiment is the outward projection of his clam-

orous need, so all he feeds upon and gathers to himself
as a possession, all that he unites with through kinship,
affinity, and the ever-broadening communion with Nat
ure and his kind, is an extension of his organism in
time and in the world, an expansion of his exhaustless litany. And all his prayers are answered. Whatever may be man's sense of responsibility,
the divine responsibility encompasses the Divine Responsibility.
universe, not only at every point unfailing,
but all-inclusive, embracing all wanderings and all the
wanderers. There is no system in which light is broken
by shadows and alternates with darkness, where the
darkness is not of divine ordinance as well as the light;
no prison-house or place of exile in which man can
ever find himself which was not prepared for him from
the foundation of the world.

IX

The Father hath, indeed, divided unto all His living.
In the structural specialisation which has gone on
with the division, one of the most striking
peculiarities is the arrest and suspense of Structural
Stability
living currents, giving things upon the earth
the appearance of stability—a tendency to solidification, to hardness, especially at points of superficial con
tact, until the hardness becomes brittleness, and from
extreme attrition all things seem to come to dust. While
this is more noticeable in inorganic matter, it is also a
characteristic of organisms. With the hardening of the
earth's crust there comes to be a tougher fibre of plant
life, and the vertebrate animal appears; and in each

individual organism age is indicated by the induration
and fragility of structure. The hands grow hard like
the things they handle, as do the soles of the feet from
walking. Use and wont beget indifference and even
cruelty in the moral nature. Institutions have the same
tendency; rituals become formal, governments rigid
and perfunctory, industry a dull routine. Social refinement at its extreme is hard enough to take a polish,
and aims to present a front of cold and staring imper-
turbability.

The points of contact between man and the outside
world, after the period of his first childlike wonder has
passed, are mainly those associated with his handling
of material things that may be moved about and manip-
ulated at his option. The timid reverence that belongs
to tender sensibility is dissipated by familiarity, which
leads first to naïve play, wherein there still remains a
trace of shyness, and then to the bold workmanship of
the artificer. The wandering stream of nomadic humanity is arrested, and the movable tent gives place to
the fixed dwelling. Social stability obtains firm foundations; the shepherd with his living flocks becomes an
episode, lingering in the fields outside the growing city;
metals, at first used only for ornament, are coined into
tokens of commercial exchange; temples are built for
the worship of Him who was once sought in every living fountain; and over the dust of kings arise the
pyramids.

All this is but a continuation of that terrestrial development by which the rock-ribbed continents emerged
from the flowing seas; and as upon the continents the
web of life is woven in more varied shapes of plant and

bird and beast, so about the fixed structures of man's making flows the human current in a slower movement, but statelier and more manifoldly beautiful. The insulation and stability are only relative ; nothing is permanently held aloof from the general circulation. Water held in the closest receptacles sooner or later finds its way to the sea ; and the sea, which is forever eroding and transposing continents, is itself continually dissolving in vapour. Resistance becomes the fulcrum of leverage. There is no point of rest in the universe.

Nevertheless the progressive specialisation of life lays stress upon the separateness and insulation, and this emphasis of Time punctuates the Word from the beginning, until that Word is made flesh in the Christ, who gathers up all the fragments that none may be lost, who shows us the Father, and who is himself utterly broken and made whole again before our eyes, that we may comprehend the glory of Death.

X

The emphasis of Time begins with Creation. Beginning is genetic, creative, on its unseen side eternal, though conceivable by the mind only as in time and space. Time, etymologically, means *The Emphasis of Time.* something cut off, a section, a season (*tempestas*); and in like manner we think of space as something in allotment. Study demands attention—an arrest of thought regarding an object also held in suspense. Thus contemplation (from the same root as time) im-

plies the intent beholding of things confined within a
circle cast about them, like a spell in magic.

All these terms, signifying confinement, definition,
arrest, suspense, are expressions of finitude, of a world
passed, as it were, from its genitive to its accusative
case—to the field of objective reality, appearing in this
view as measurable matter and motion, as broken in
time and space into related parts and sections and even
into inert particular fragments that often, though nearest our hands and feet and emphatically real, seem
irrelevant, trivial, and inconsequent.

It is far away from the plenty of the Father's house
to the husks on which men starve. There the abounding, eternal life—here the limitation and involvement;
there the infinite power—and here at the end of things
mere dust and impotence, empty travail, stumblings,
vexation, defeat. Finally the extremes meet—starvation and the feast, sickness and healing. Not a sparrow
falls to the ground without the Father's notice, and the
very hairs of our head are numbered.

We must needs continually keep this everlasting
nearness of home at heart, and in this personal way,
because of the apparent remoteness, and because the
ordinary course of thought as well as the tendency of
scientific analysis is toward an impersonal view.

XI

The divided living or, as it is scientifically phrased,
the specialisation of life, is a development stretching
through long periods, each of which is marked by the

appearance of some new form of existence upon the
earth. While the older theology accounted
for each new stage of development by a ^{Creative Spe-}
series of special creations, modern science
has sought to exclude altogether the idea of any creation, regarding each new form of life as evolved from
antecedent forms through natural selection and the
modification of environment. The older theology, as
represented by Paley, attempted to explain the marvellous adaptations which constitute the rhythmic harmony of the universe by the operations of an intelligence patterned after the limited and specialised human
understanding, first choosing to create and then arbitrarily choosing means for the accomplishment of rationally conceived ends. Modern science has, in the
rejection of this idea, gone to the extreme of repudiating divine purpose, explaining cosmic co-ordination
by an impersonal selective wisdom inherent in matter
itself.

By substituting creative specialisation for special
creations and postulating a supreme Personal Will
and Intelligence, transcending specialisation
but immanent therein, with a purposiveness
spontaneous in its working, not according
to plan as the result of choice (in our human sense of
the term), but showing a plan, not limited by alternative, but itself the ground of alternation, Christian philosophy presents to science not merely the ground of
common agreement, but a view involving no more mystical assumption than is involved in the postulations
made by science itself of an invisible ether and an invisible atom —whether the latter be considered the ulti-

mate material particle or a vortical motion of the ether
—a view, moreover, which, accordant with faith in a
loving Father as the source of all life, also clears the
scientific field of problems that from their very nature
are insoluble by any possibly discoverable facts. Such
problems as are presented in the questions: What is
the origin of organic life upon the earth? and How is
the psychical developed from the physical? cannot be
solved by any data lying within the limits of scientific
investigation; they arise, indeed, and assume their most
formidable shape through lack of faith in the sufficiency
of creative life for its own transformations. There is
really no greater chasm between the inorganic and the
organic, between neurosis and psychosis, than there is
at any stage of the progressive specialisation.

Regarding specialisation as at every point a creative
act, the problems disappear; Life itself becomes the
great bridge-builder—the *pontifex maximus;* and considering, furthermore, that reaction proper to Life, whereby it has solvency and escape from any individual synthesis, and even in due time repents itself of an entire
species, rising again in some other body or in some
wholly new type of embodiment, then indeed that singular Voice, saying *I am the Resurrection and the Life,*
may give the transcendent note to philosophy as it does
to faith.

It is indeed a singular Voice, and proclaims a truth
which, from the foundation of the world, has been
hidden.

XII

The progressive specialisation of life is not through evolution primarily, but through involution, every new stage of progress being a new folding of the veil. The universe is not an unfolding of God, but a folding of Him away from Himself, until the manifold hiding is completed in the human consciousness which is the ultimate fold of all. There could be no more arbitrary and mechanical conception of God than that of Him as a vast involute, implicating the universe. Progress would then be from what is most complex, through a series of explications to what is most simple. This is pantheism in its baldest form, abrogating the mystery of Creation, which is also abrogated in the theory of emanation.

Both the transcendency and the immanence of creative life begin to be hidden with the beginning of existence. When God said Let there be Light, the light became the first veil hiding Him. Therefore it is that no man hath seen the Father at any time, and when we speak of His power and wisdom and purpose we have no real idea of these attributes, which are known to us only as mediate and limited ; and when we say that He is Love, we express not the reality, since our knowledge of love like that of light is from broken images only. From the beginning, then, is the eternal life hidden, and though the veil hiding it be light itself, that which is concealed is beyond our expression in thought or speech.

At every successive stage of the cosmic develop-

The Hiding of God.

ment, or rather envelopment, there seems to be a fresh
surrender of potence and sentience, though with greater truth it may be said that these are more and more
veiled ; the organic tending toward variability and fallibility as compared with the inorganic, and the vegetable instinct being surer than the animal. With the
growing complexity there is increased uncertainty and
indirection, until we reach the hesitancy and vacillation
of rational volition.

As heat is given up in the contraction of the earth
and the incrustation of its surface, and as the solar fire
is subdued to a lambent flame which runs through all
the variegated terrestrial life, so is the universal pulse
modulated more and more down to its measured beat
in the animal ; and with the increase of temperament,
adaptation considered as correlation, and outward correspondence, interaction and interdependence are more
pronounced, just as with the loss of heat there is greater conductivity.   Sentience, which is really mightier in
the less specialised forms of life, yet appears outwardly and in definite expression more intense and finer
in more complex forms, and is more communicable
mediately as it is the more patent. With the veiling
comes gain as well as loss, so that we properly think
Gain of Per- of later forms as more advanced, though in
spective in a certain sense division signifies diminution.
Specialisation. Sight is possible because the eye is a re-
fractive lens, and thus only because the solar light is
tempered by the medium of an atmosphere.   The
blind feeling of which sight is a specialisation is a
surer sentience, a divination knowing no distance or
indirection, a wisdom therefore whose ways are never

missed, an unbroken clairvoyance ; but the intense, confined specialised sense of ocular vision is an outward openness, aware of expansion. That which was once blind, feeling its way by dead-reckoning, like a mole in the dark, now takes in the heavenly blue and (under the veil of night) the stars beyond, making for itself a wondrous perspective. The same blind feeling specialised as hearing catches vibrations slower than those of light, making for itself another perspective of beautiful harmony.

XIII

It is not unity which is divided ; our conception of unity is the reflex of our thought of the manifold. It is not identity which is diversified. Absolute homogeneity as the initiative of a universe is the most sterile mental notion ever conceived in the attempt of philosophy to evade the mystery of Creation. From such homogeneity there is no genetic thoroughfare — no way out. The idea of absolute heterogeneity, on the other hand, leads only to chaotic distraction, which has no recourse, no reaction, no way back reflexively into the consistency of a universe. The genesis itself—a mystery hidden from human comprehension, yet mystically apprehended— is action and reaction, and we see that as a manifestation it involves at once the idea of otherness and consubstantiality, the nearness and kinship, at every remove, being the reflex of distance. We have a mechanical and therefore inadequate illustration of this action and reaction, as essentially one and inseparable,

Tropic Reaction.

in all tropical movement. Thus the earth in her flight
from the sun is, at every moment of the flight, returning; as in her rotation her turning from the sun is at
every point a turning to him. The reaction is in the
action, and we cannot logically separate the one from
the other; and when we separate them historically, we
make the diversification primary, supposing flight to
precede return, repulsion attraction, and all functioning its inhibition. Following this rule of precedence,
we should reverse the procedure of Herbert Spencer's
synthetic philosophy and give the initiative to heterogeneity.

XIV

Certainly it is the reflex, the feeling of the ancient
bond of nearness, that is more and more hidden from
the planetary prodigal in that far-country perspective
which at every step becomes more bewilder-
Surprise and
Familiarity. ing in its varied charm of beauty and delight. There is no new surprise which has
not in it some homely reminiscence, but, by the very
urgency of destiny, it is the surprise itself which cap-
tivates and absorbs, leading the wanderer further
afield.

XV

There is this preoccupation and expectancy in what
we call the inorganic world. Here, at some critical
point, there is a sudden departure from the uniform
succession of phenomena, and a new synthesis is apparent, not explicable by any antecedent situation or by

any elements visibly entering into the first combination. These transformations do not come about according to such laws of causation as are formulated by the human mind for the explanation of phenomena that come within the range of conscious volition. By no calculation based upon any deductions of science could these surprising changes have been anticipated. " If we conceive," says Mr. N. S. Shaler,* " an intelligent being looking upon a mass of nebulous matter having only those forms of associa- Surprise and Prophecy. tion which are possible in gases, we must believe that such a being would have been entirely unable, if his intelligence were less than infinite, to form any conception of the results which would arise when that matter came to take the present shape of this earth." But that intelligence which is immanent in these transformations is prophetically expectant, seeing the end from the beginning. It is not, therefore, to be supposed that to this profound intelligence there is no delight in the wonderful surprises of the ever-changing world. To man also belongs a prophetic vision, however hidden or obscured, looking inerrantly toward the Things to Come, and it is because of this undefined and sure expectation, and not because of anything outwardly seen in the novel wonders of progressive life, that he has delight in them ; while a life that in all its changing scenes should be the exact fulfilment of definite mental anticipation would, on the other hand, be tiresome, not answering to the unseen hope.

Nor is there such uniformity of routine even in the

* *The Interpretation of Nature*, p. 55.

succession of grand cycles as to mar the divine delight in creation. If at the end of each of these cycles the entire universe is dissolved, the synthesis of a new universe would not be the exact repetition of that which preceded it. No cycle of life returns into itself; the death completing it is always a transformation.

XVI

In the progressive cosmic involvement the reality of Life—in its essential attributes—seems to be more and more hidden beneath' appearance. Form hides the formative; and form, persistent and held in apparent suspense, veils transformation. The appearance of uniformity in the physical world especially impresses our human intelligence, which is confined in its observation and investigation to a very limited period of cosmic development—the period of greatest apparent stability. If our point of view
Uniformity Veiling Trans- formation.
were transferred to an epoch indefinitely remote, before the appearance of organic life, while we would have the sense of order, yet would uniformity seem a transparent veil. We were not indeed shut out from that simplicity; rather are we shut into the present manifold complexity. The quality of life is the same, whatever the situation ; and its manifestation in the simplest form was a Habit, however loose and flowing in its longer waves of pulsation and its marvellously swift alternations — its appearings and vanishings. When he who is eternally the representative of man spoke of sharing with him the glory of the Father be-

fore ever the world was, he was speaking of our native
heritage. The divine nature never had a habit that
was not also human, and the man we know—the ultimate creative manifestation—still reflects that nature,
as its very image.

That period of comparative simplicity, when the
world which we call inorganic and lifeless was the
only living world, was vast as the ocean when measured against the mere island in time which is occupied by animate existence, as we know it, in all its
wondrous variety. Now are we sheathed in integu-
ments that hide the older world which we still unwit-
tingly inhabit. When the darkness of our little night
lays open to our eyes the starry spaces we may still
behold the plasmic milky-way of that long night of
time whose possibilities were mightier than we can
even dream in such sleep as now befalls us.

What we know as desire is away from all this, projecting its embodiments into that narrow island of
specialised life which, after all, still rests upon that
hidden ocean. We bask in what seems to us the
nearer and more familiar sunshine, and turning from
that simple estate which we still hold in the darkness
and which still holds us—that older deep which ever
felt the brooding Spirit of Life — we rejoice in the
broken lights and casual acquaintances, in the color
and temperament, in the poise and modulation of a
suspended world. We glory in difference as a distinction and in individual isolation as the proper integrity of an organism, its inviolable virtue.

To us, in the suspense of a fixed order, the processes of Nature seem to be movements in cycles that re-

turn into themselves. "The sun ariseth, and the sun
goeth down," says the Preacher, "and hasteth to the
place where he ariseth. The wind goeth toward the
South and turneth about unto the North; it turneth
about continually in its course, and the wind returneth
again to its circuits. All the rivers run into the sea....
Unto the place whither the rivers go thither they go
again." Science, especially in the departments of
physics and chemistry, easily regards these closed
circuits as somehow independent of and isolated from
creative action and reaction. The physical world is
thus considered as inanimate, having neither life nor
death but only motion proceeding from inherent forces
and according to laws of its own. So completely is
the universe separated from a personal creator that
even those who believe that it was originally His
creation accept the illusion of delegated and secondary forces that are like the servants of the vineyard
abandoned by its master, who has gone abroad, but
who may at his option return and in some marvellous
way assert his dominion.

But in reality God is always in His world, and always working the great miracle of creation.

XVII

"No man hath seen the Father at any time," but
the Son shall reveal him. The appearance

Organic Life,
a Pre-Messi-
anic Revela-
tion.

of organic life upon the earth was the fulfilment of what may without irreverence be
called the Messianic expectation of Nature. It was indeed a miraculous conception of the

Spirit of Life, and was not without wonderful preparation and prophetic adumbration.

The later and closer scrutiny of the processes of the mineral kingdom show that in many respects these are not so sharply distinguished from those of living organisms as was formerly supposed. " It has been found that finely divided particles of many substances when suspended in a fluid will, under the influence of some forces as yet not well understood, take on an incessant movement. So perfectly does this motion resemble that of some of the microscopic forms of the lower simple organisms that naturalists at first supposed that in observing these movements they were dealing with living beings. The crystals of the rocks perform functions which were once supposed to be peculiar to animals and plants ; they undergo changes in their constitution, often taking in new materials, which they sometimes decompose into their elements and rebuild in the new growth. So, too, crystals are in a way capable of multiplying themselves, for when one begins to form, others of the same species, as it were, sprout from it, much in the manner of certain lowly forms which are certainly alive." •

After millions of years of cosmic preparation the cell appears — the precious nursling of the ages. Yet, if a human intelligence could be supposed to have been present in the world before this remarkable advent, it would have been unable to mentally conjecture what was about to emerge from the matrix of a world that seemed already so old and barren , nor indeed would

* *The Interpretation of Nature*, by N. S. Shaler, pp. 111. 112.

such an intelligence, brought face to face with this long-expected child of Time, so lowly at its birth, wrapped, as it were, in swaddling-clothes and ignominiously stalled, have had any prescience of its mighty meaning and mission.

Nevertheless this was, as we now can see, a new creation, a transformation so wonderful that only because thereof does the world seem to us to be alive. The Prince was already within the portals of the Palace of the Sleeping Earth, and the heart of the virgin planet was stirred by a new dream—the vision of a lord to come who was older than the Sun. The Sun also knew, for he was the flaming Witness of the Spirit of Life, Who was now to begin His earthly ministration with mighty miracles, turning water into wine and wine into blood. "He must increase, but I shall decrease."

Already, indeed, from the beginning of cosmic specialisation, there had been the diminution and descent —the macrocosmic yielding to the microcosmic mystery, the whole magnificent universe narrowing its circles, contracting its spheres, veiling its potencies and lessening its velocities, stooping down to serve the coming Prince of a new kingdom, all its strong, wise ministrants gathering at his nativity, like worshipping Magi, bringing special gifts, also, like the gold and myrrh and frankincense of the Eastern Kings, signifying Abundance, Burial, and Ascension.

XVIII

The cell is not the introduction of life into a dead world. The universe was from the first living and sentient in its macrocosmic order, organic in that order. The term inorganic is not properly applicable to what was from the first an organism and constantly reaching forward to more complex organisation. Nevertheless the cell marks a pivotal and critical point in the progress of existence. Its appearance is a surprise, a fresh embodiment of the all-shaping Power and Wisdom; but there is nothing more mysterious in a germ that grows than in a mineral which crystallises; it is the old mystery in a new shape.

The Old Mystery in a New Form.

The new integration is not explicable through what precedes it; it would be truer to say that it is the explication of all its antecedents. It is itself a new hiding of life, a fresh strain of cosmic tension, a further division and suspension, a more discrete modulation, a more exquisite temperament. The outward balance of things, already so nicely adjusted, maintained through oppositions and contradictions, through attractions and repulsions, through ascents and descents, may have been suddenly disturbed through some vast dissolution of existing forms, liberating mighty forces for a new continence, and so have regained equilibration by the storage of this precious argosy, freshly launched upon the ocean of existence. But, even so, we are only attempting to express the mystery in the terms of an outward equation, what is lost on the one

side being gained on the other, as the ascent of one
arm of the scale is a descent of the other.   Dissipation of energy is the concomitant of all integration ;
but these terms are not related to each other as cause
and effect (any more than one-half of a circle is the
cause or effect of the other); they are merely complementary.   The specialisation is creative.

The evolutionist, while he helps us to see what is the
true outward sequence, confesses his inability to show
causation in the sequence.   "The ultimate mystery—
the association of vital properties with the enormously
complex chemical compound known as protoplasm—
remains unsolved.   Why the substance protoplasm
should manifest sundry properties which are not manifested by any of its constituent substances, we do not
know ; and very likely we shall never know.   But
whether the mystery be forever insoluble or not, it can
in no wise be regarded as a solitary mystery.   It is
equally mysterious that starch or sugar or alcohol should
manifest properties not displayed by their elements—
oxygen, hydrogen, and carbon—when uncombined.   It is
equally mysterious that a silvery metal and a suffocat-
ing gas should by their union become transformed
into table-salt.   Yet, however mysterious, the fact remains that one result of every chemical synthesis is
the manifestation of a new set of properties.   The
case of living matter or protoplasm is in no wise exceptional." *

* John Fiske's *Outlines of Cosmic Philosophy*, vol. i., p. 434.

XIX

This protoplasm is the nebulous beginning of what
to us seems like a distinct universe, peculiarly open to
our sympathetic comprehension because of
its intimate association with our earthly fortunes and destiny, since humanity is its ultimate issue and fruition.   Physical and chemical processes seem remote and obscure save as they
come into immediate contact with our life : in the air we
breathe, the water we drink, and the component elements of the food we eat; in the minerals which lend
themselves to our use in various ways , and in the light
and heat and electricity which seem like a part of our
vitality, and which outwardly are elements of comfort
or disturbance, conservation or destruction, according
to their temperament.   And beneath these is the universal physical bond of gravitation, which enters into the
rising wave of life, in certain forms of attraction, as a
ladder of ascent, and in the falling as a lethal burden.
Modern science has given us a clearer idea of these
forces and elements in their quantitative relations, and
a wider and more effective adaptation of them to our
use ; has made of their rhythmic motions a fairy tale
for wonder, a beautiful poem; has shown how the
world has given harbourage to vegetable and animal
life, within what narrow limits of temperature is possible the chemical action upon which molecular organisation depends, and within what still narrower limits
a physiological synthesis can be maintained ; how elements like oxygen and hydrogen, which at a high degree

of temperature combine to form watery vapor, may at a
lower degree rest side by side for independent action,
and how peculiarly essential is nitrogen to the storage of
animal heat : in all these ways indicating the care and
providence of a loving Father in preparing a dwelling-
place for man. But we are so involved in the organic
synthesis that we translate all physical terms into those
which are more intimately familiar to us through our
specialised physiological sensibility and mental percep-
tions, our language in its primary meanings leaning
rather to the former, and, in its secondary, to the
latter.

In the period of naïve impressionism the whole universe was humanised, and even the gods were included
in this general incarnation ; and, considered simply as
to its reality, this impression was profoundly wise—a
deeper divination than the human reason reaches in
its supersensuous mathematics and formal knowledge,
though these have more truth of perspective and a
more exact discrimination. The extreme rationalistic
view of the world excludes all humor from its dry light ;
reduces the sensibility to the humble offices of a ser-
vant to the intellect—otherwise burying it out of sight ;
and rejects physiological and anthropomorphic interpretation. This is the inevitable tendency of specialisation, which at every step is a new veiling of life—
of its essential wisdom as of its potency. The truth of
chemistry is not the truth of physiology, as is shown
by the inadequacy of the chemical analysis of food as
a test of its physiological action. So the truth of
physiology is not that of psychology. We see, then,
how remote and alien the true and proper life of what

we call the inorganic world must be from our mental vision and even from our sensibility.

XX

We are so accustomed to regard different forms of life as higher or lower according to their place in a progressive series, and thus to unduly emphasise the superlative importance of the most specialised Higher and existence, that our view is distorted. In Lower Life. this way we come to depreciate the living values of pre-human nature. We form the same comparative estimate of different periods of human history, underrating the eras of greatest simplicity, and in like manner, considering an individual life, we attribute a superior excellence to maturity, as if we should prefer August to May. Consistently with such judgment, we might reasonably question why manhood is not sustained at its ascendant; why one generation should pass away and another come, repeating the crudeness of infancy; why the sun is not maintained at the zenith; why civilisations disappear; and why, indeed, all systems are doomed to dissolution. Reversing our pref erence in any of these cases, our view would have the same fault of disproportion.

Our human conduct, under the extreme limitation of arbitrary and fallible choice, is so much a matter of experimentation and discipline, involving moral preference, wherein rising is a betterment and falling a vilification, and having for its ideal field some lofty pla teau of stable and perfect goodness, unmixed with evil

7

and undisturbed by reactions, that we come to regard the progression of all life as having this moral character, as if the Creator were in the same toils of experimentation, learning to create, and improving with each new creation. To rid ourselves of this illusion, whereby our limitations are transferred to the Infinite, we have only to see that while there is always the world to come, it is not a better world, according to moral preference, but a new world; that the creative life repents of the good grown old as well as of the inveterate ill; that this life is in its essential quality a transforming, regenerative life.

The vital perspective is that of a circle, wherein compensation is everywhere apparent—not a circle returning into itself, but involving endless permutation and variability. We need not resort to the familiar similitude of a spiral ascent. To the undisturbed spiritual insight there is no higher or lower, no superiority of the molecular to the molar, of the chemical to the physical, of the physiological to the chemical, or of the psychical to the physiological. As has been already said, the quality of life is the same whatever the situation.

When we think about nebulous expansion, ethereal vibration, molar or molecular attractions and repulsions, our thought is empty as compared with our sympathetic apprehension of those actions and passions which belong to what we call the realm of biology.

XXI

Our vision of life is like that of Jacob at Bethel, one of ascending and descending angels; but the angels descending are the same angels that ascend. If the world were only "inorganic," and such only it is believed to have been through the greater part of its existence, it would still have all the excellence of life in its essential quality —an ineffable excellence of which we have no conception. Its ascending angels for the most part elude our vision, only its descendent ministration being apparent to us. The side of the inorganic world presented to the organic is the dying side—chemical dissolution next to physiological integration. The crescent organism confronts a world which is dying that it may live. The cosmic accommodations which have made the earth man's dwelling-place have been renunciations of life in his behalf; and the dead moon is a nightly reminder of that Calvary from which Nature stretches forth to us her skeleton hands and shows us her hard, dumb countenance. When our Newton comes, it is in the autumn field that he finds in a falling apple the suggestion of the universal law—that of gravitation, the symbol of death.

The sun himself is dying, giving forth his light and heat; he is a true martyr— the witness of the Lord; and the coal deposits buried in our earth ages ago are like the famed ossuaries of martyrs, having stored-up virtues for miracles of warmth and light and healing.

We love to dwell upon this descent of the Lord and

Why the Inorganic seems Dead

his angels in the world of inorganic matter, which we
call dead ; in the light and heat and the refreshing
rain ; in the virtues of the cooling earth ; in chemical
disintegrations, and to see that it is all a descending
ministration for the lifting up of organisms. It is a
view of the world which invests with our pathetic
affection its very débris and the dust we tread upon.

Nature, in our observation of her apparently closed
circuits, is known to us, outside of organisms, mainly
in her descents for the risings of these. What are her
own proper ascensions for this beneficent ruin, or
what is her own World to Come—her transformation,
answering to our Resurrection—is hidden from us.

Biology, notwithstanding its rigid exclusion of the
inorganic world from its proper scope, furnishes suggestions for the poetic and spiritual rehabilitation of
that world in the human imagination,

> "And in our life alone doth Nature live,
>   Ours is her wedding garment, ours her shroud."

Modern sentimentalism has undoubtedly carried this
ideal rehabilitation to an extreme, transferring to Nature solicitudes wholly alien to her and purely human,
and needing, therefore, for its correction, the scientific
comprehension of what is peculiar to physiological and
psychical specialisation.

XXII

The cell germ is the central sun of the physiological
planetary system — the beginning of a new career

of prodigal wandering. The earliest and simplest organisms are unicellular, as if a new kind of universe
were begun in a single-mansioned econ-
omy. But what singular potency in this
Special a-
tion of Sex
and Death
simplicity! This is shown in the ease and
quickness of reparation, by which any part of the organism lost or destroyed is restored. If the body
is cut in twain, each part continues its independent
life; or rather we should say that such separation has
not at this stage of development the meaning which it
comes to have when the organism becomes more complex, consisting of interdependent members. While
identity seems to be emphasised, yet there is the tendency to diversification. Reproduction is by division,
by simple fission. In the *infusoria* reproduction is
preceded by a comatose state resembling death, an
arrest of activity during which the identity of the
parts soon to be separated seems to be assuredly es
tablished; and after the fission there is no distinction
by which one part may be designated as the parent
rather than the other. Such organisms, as Dr. Weiss
mann has shown, have a kind of immortality, suffering
death only as an accident. The amœba of to-day is
the original amœba.

With the multiplication and diversification of cells in
later and more specialised organisms, there is allotment
of function, a division of labor and an interdependence
of co-ordinate parts, and the same appearance of delegated powers which is characteristic of complex econ
omies, just as there seem to be secondary forces in the
inorganic world, which we think of as acting independently and yet interdependently as mutually related.

With the specialisation of sex—a divulsion for union, a repulsion for attraction—death also appears as a specialisation, entering the world hand in hand with love. From this point the variation goes on with remarkable rapidity.

XXIII

The appearance of organic life upon the earth as a prelusive analogue of the appearance of the Christ-life in the human cycle has already been suggested. It is thus seen to be one of the successive revelations of the creative Logos. The analogy would require a separate thesis for its full elaboration. It is only important here that we should draw attention to a few points touching our present theme.

The Gospel of the Cell.

1. The organic involution is the apparent beginning of a motion of return. It is the beginning of the disclosure of conscious life, reflecting Godward. This attitude of the vegetable and animal kingdoms was recognised by Swedenborg.

2. The organic plasma, having its matrix in an apparently dead world, is the beginning of life in a procession of generations. It is the physical analogue of that childhood which is the type of the Christ-life.

3. Cell-life, in its simplest and most plastic forms, has a marvellous potential energy, with spontaneous power of self-reparation, and thus foreshadows miracle-working and redemption.

4. The organism grows, and is thus the physical sym-

bol of the increase and authority of the "more abundant" Christ-life.

5. The most significant point of the analogy is the concurrent specialisation of sex and death : that with the love which is the basis of genetic kinship came a new mortality, just as in the spiritual development of humanity the love which was the ground of a divine-human fellowship was bound up with a divine-human death.

These points, more fully dwelt upon hereafter, are here brought together as a natural introduction to the consideration of the organic movement toward incarnation.

XXIV

Certain aspects of life, elsewhere hidden, are visibly revealed or suggested in the realm of physiological activity. For while the embodiments of this realm are veils hiding life, and are indeed more complex, a closer network and imprisonment, than are the manifestations of physical and chemical energy, they are at the same time more open to our study and *Physiological Intimacies* comprehension. Their history is more recent, and has left its traces in fossil structures embedded in the rocks. Many of the earliest species of organic life remain in living specimens upon the surface of the earth for our observation of function as well as of structure. Moreover, to us as a part of this realm —its ultimate issue and consummation —there are intimate disclosures of its processes in our own sensibility and consciousness. We know what desire is and aversion, hope and fear, pleasure and pain, action and

passion, faculty and capacity, aspiration and depression, sympathy and conflict, confinement and release,
rest and disturbance, bounty and want, demand and
renunciation; and we know that structure is for these,
and not these for structure.   We breathe and eat and
sleep and love and die; and we have a sense of our
incarnate action and passion as such, and apart from
considerations transcending physiological limitations.
Beneath these is the eternal ground of these, the Word
which becomes flesh, and which in the flesh has again
its glorious appearance as the Word; but we are now
considering what revelation of life there is in the wonderful organic development of incarnation itself, excluding from the scope of our contemplation that
specialised intelligence which distinguishes man from
other animals.

The transformations by which the " inorganic " world
has come into its present state are hidden from us,
whereas in the historic development of organisms the
series, though not present to our view in its completeness, is to such an extent observable or indicated as to
be profoundly impressive.   Apart from the historic
series, there is everywhere open to our observation a
miracle of growing life which directly suggests the creative power.   Nature becomes to us a Book of Genesis.
We seem everywhere to hear that first of all commandments, Be fruitful and multiply.   And in this book of
Genesis how inevitably does the mind pass from the
first chapter, in which the earth brings forth the herb
yielding seed and the fruit-tree of every kind, whose
seed is in itself, and every living creature, multiplying
its kind, to the second chapter, in which it is declared

that all these were created before their appearance
the plant "before it was in the earth and every herb of
the field before it grew." The growth is the outward
manifestation of that genetic quality which is the eternal
attribute of boundless and abounding life.

XXV

We have seen that death, as a specialisation, enters
the world with love. There is an adumbration of this
association in the nearness of all desire to a kind of
death. Nutrition is the rising of one wave
next to the subsidence of some other, and   Nutrit n
the wave that rises is not the same wave that falls.
Growth is genetic transformation. This nutrition is
one of the most suggestive of the object-lessons furnished by organic life. The nucleus of a germ is first
manifest as a living thing in feeding upon its envel-
oping substance or integument. In the case of a seed,
so long as the outward muniment about it is secure
from dissolution, its power is latent ; but being buried
in the earth, where outwardly it is in peril, it inwardly
escapes, is liberated from its imprisonment, and feeds
upon its crumbling prison-walls. The nourishment
thus begun is in the same way extended. The flame
once kindled upon the altar spreads, devouring substances beyond its original source of alimentation.
The vegetable, rooted in the earth, feeds upon the ele
ments that come to it, these being broken for it, dis
solving for its integrity. The animal carries its roots
about with it, having voluntary locomotion, and in its

wider range of selection compels its victims. The descent of the inorganic is for the rising of the vegetable,
which, transforming the material for its subsistence
from the earth and air, becomes itself a broken sacrifice for a new transubstantiation, falling for the rising
of the animal. From the first appearance of a cell to
the advent of man stretch millions of years, and at his
appearance the world has become his pasture, through
numberless varieties of vegetable and animal life. The
Lord is the shepherd. There has been this shepherd-
ing from the beginning of organic existence, life feeding upon broken life. The functioning of organs thus
nourished is a wave of motion rising next to the dissolution of these members. And there are waves
beyond these, not properly within the scope of our
present consideration — the continuation of the descending ministration, until the Lord becomes the
shepherd of souls—always a dying Lord.

XXVI

In the inorganic world we more especially note the
division involved in specialisation and the progressive
diminution, as in the contraction of spheres—the gravi-
tational contraction of the sun being itself
a depression, or descent, for the generation
of the heat and light of the planetary system.
But in organic specialisation we see the division as
more conspicuously a multiplication—an increase. The
abundance of life is visibly manifest.

The vast amount of heat generated by the contraction

*Organic Re-
version of the
Inorganic.*

of the sun must be very much diminished before organic
life is possible upon the surface of the earth. But in the
progression of organic life the store of heat is continually increased. The earliest animals are cold-blooded.
While the processes of the inorganic world tend toward
an appearance of rigid uniformity and fixed stability,
those of the organic render more conspicuous the appearance of variation, and the more complex the organism the greater becomes its instability ; and in many
ways the procession of organisms seems to reverse that
of inorganic matter, though in reality it only makes
visible to us tendencies and attributes of life which in
the macrocosmic procession are hidden from us. This
visible manifestion begins, indeed, for us in molecular
organisation as shown in the field of purely chemical
action. Thus the mineral, water, in its various states,
solid, liquid, and gaseous, more than adumbrates the
suggestions received by us from physiological action.

XXVII

The distinction we have noted seems to be rather
between the molecular and the molar than between
the organic and inorganic synthesis ; and this distinction would doubtless disappear through a more intimate acquaintance with the molar universe. The
atoms of a molecule imitate the motions of
the solar system—having attraction and repulsion and tropic movement, dissociation
and reassociation of the dissociated atoms. The
study of solutions, combined with that of thermo

dynamics, and later with that of electro-dynamics, has
thrown much light upon the vexed problem of the constitution of matter.   But even the simplest observa-
tions regarding so common a substance as water
comprise phenomena that look back to primordial
embodiments of mist and flame, and forward to the
flame of life incarnate.   In its gaseous form, water is
absorbent of heat, which at the same time expands
and lifts it, and yet with this expansion there is a tension, as within the limits or bounds of its capacity, a
confinement by invisible walls.   Or, to express the phenomenon in another way, the heat expanding the air
makes it an absorbent of water, so that the flame has
an embodiment of vapour, both the embodiment and
its confines becoming invisible ; and this expansion
goes on until the tension reaches its limit of capacity,
when at a critical moment there is the explosion and
precipitation—the descendent ministration.   We have
here a prophecy of the latency and storage of energy
in physiological capacity, as when the flaming desire
shapes the mouth of an animal, expanding it inwardly
into a stomach as a receptacle for food, and into the
lungs as a receptacle for air.   As these organic capacities are deepened inwardly, representing in their
sphering and involution and convolution the synthetic action of cosmic envelopment from the beginning, the desire which has thus shaped itself by intus-
susception, expressing its postulation, is outwardly a
flame of increase, ascending also while it is crescent
until it reaches the culminant point of its physiological term, where it quickens and flowers and falls.
Water, when by the dissipation of its latent heat it

reaches a certain critical point, suddenly quickens, and, instead of contracting, expands into its florescence of crystallisation, here again foreshadowing that epoch of organic development which determines generational succession, where the flame of increase becomes for it own organism a consuming flame of sacrifice, falling to rise again in another but consubstantial incarnation. We shall consider this point more at length when we come to treat in greater particularity the ascent and descent of life. We wish here only to draw attention to the fact that while increase is so conspicuous in organic existence, death is equally conspicuous, and is thus emphasised at the very point where nutrition is arrested and transformed into a genetic process.

Death, which invisibly is Love—the attraction of gravitation in the spiritual as in the physical world, binding all spirits to the Father of Spirits, as all planets to their suns, and bringing all prodigals home—is also born of Love, when it visibly and conspicuously appears as a specialisation, in connection with the procession of generations in the organic kingdom.

XXVIII

Resuming the suggestions derived from a study of organic specialisation, we find that they contradict certain propositions which are accepted as axiomatic truths in the realm of physical science : or, rather, they introduce opposite propositions essential to a full comprehension of Nature, of which science professes to give but one side.

1. Science, dealing only with structure and function, lays stress upon evolution   A philosophic view of Life as transcending structure, as creative, brings into prominence the opposite truth of Involution.   In a single passage of his Synthetic Philosophy, Herbert Spencer admits that this philosophy would be more truly indicated by the term Involution;* but generally his consideration of nature ignores not only creation but Life itself, and is confined to sequences so stated as to imply the evolution of every new form of existence from its antecedent.   In reality, the term evolution is properly applicable only to the processes of expenditure, ignoring the original tension.   It is as if we were to consider a watch wholly with reference to its function as a time-keeper — an office which it performs through the relaxation of the tension of its spring— giving no adequate consideration to the tension itself, because our attention is fixed upon the action of the escapement as more immediately associated with the use or function of the machine.   The scientific man does not ignore latent potency, about which, indeed, he has much to say.   He will show us that the potential energy of the sun is greatest when its distance from the earth is greatest, and when, therefore, the kinetic, or patent, energy between the earth and the sun is least ; but it is energy as kinetic, as manifest motion, that comes within the scope of his measurement, and whose laws he can formulate : the potential energy, on the other hand, he does not ignore, but simply assumes as the X, or unknown and indeterminable element in

* *First Principles*, p. 265.

his computation, treating it as wholly divorced from creative life, since his proper business is with motion, not with creation.

II. The axiom that motion is always in the lines of least resistance, while it is true of motion as functioning, is not true of the lifting power of life which gives tension. Of motion before it moves, if we may be allowed the use of such an expression, the opposite proposition is true—namely, that it seeks difficulty. Life as creative, as genetic, as in its specialisation a series of transformations, withdraws from the facility of habit, of a descending motion, for new involution. Moreover, this tendency of life toward difficulty rather than toward facility is illustrated in the continuation of the same species through the procession of new generations.

III. Modern scientific views, as generally accepted, lay undue stress upon the struggle for existence as a competition between species and between individuals of the same species. The result of this conflict is expressed in the familiar phrase, "The survival of the fittest." Since structure itself is for stability and conservation, within the limitations imposed by life itself (*i. e.*, by the special form of life), it is true that, other things being equal, the structure which is best adapted to its environment will have the greatest stability. There is travail in all forms of life, the struggle for a foothold, the competition for vantage-ground. As has been already remarked, life seeks difficulty, and the progress of specialisation involves at every stage of increasing complexity greater difficulty and more frequent and varied risk. An exceptionally fortunate environment

leads more often to degeneration than to the promo-
tion of fitness.   The suppleness of the pursuer is not
more remarkable than that which is developed in the
game pursued.   Taking the widest range of observation, we do not find that either safety or ease is an ultimate objective aim in Nature ; she emphasises discontinuity rather than continuity, revival rather than
survival, running toward death in her progression,
burning all bridges behind her as she advances.   In the
largest view, stability is an illusion, uniformity a disguise, the persistence of type not an eternal concern.
Life, comprehending all involvements and the solicitudes pertaining to these, has itself no solicitude, and,
because it is essentially resurrection, it glorifies death.

The term survival is merely relative, and the conflict
for survival is a part of the universal harmony which,
in the partial vision, it seems to contradict.   When we
consider that organic existence is possible only because
of a descending ministration from the beginning of a
cosmic order, and that it is sustained only through the
continuance of this ministration still further expanded
in the relations between the various species of organisms, and in the succession of generations, we comprehend that sacrifice is as conspicuous in the natural
world as is demand, that there is no cycle of existence
in which altruism is not as fully illustrated as individuation, interdependence as independence—this illustration becoming more luminous with the progression of
organic life.

Science in its specialisation deals with matter as
habit-taking.   As morphology it considers the habit as

one of structural formation. Considering the habit as
one of functional activity, it formulates the laws of this
activity which in the organic world are called the laws
of physiology. In either case the habit is an investiture, and as an outward visible manifestation hides the
principle of its own Becoming. The creative life thus
veiled must forever remain a mystery. Looking toward
the beginnings, seeing in every moment a renascence,
we find the veiling a revealing. There is even thus an
illusion, but the veil is at least transparent. But in the
study of an order we regard mainly the meanings of existence with reference to outward ends ; we follow the
stream away from its fountain ; we are lost in these divergent paths, and what we see of life appears to contradict the essential quality of life. Science in its very
modesty, in the recognition of its limitations, tends to
agnosticism. What at first was inevitably an illusion
becomes a delusion. The transcendency of life is not
apparent in the confinement of closed circuits, and its
veil is no longer a transparency, but an obscuration.
What began in modesty may thus end in inflexible certitudes.

The habit of life has been truly and patiently followed into its most intricate folds, but the scientific
prodigal has gone into the far country with his particular share of the Father's divided living ; and to him,
with his face turned that way, the order of things which
is the subject of his close scrutiny is seen true, but in
those aspects which contradict its essential truth. The
propositions which he makes concerning this order of
things, such propositions as those we have been considering, are verified by all the facts within his range

8

of observation.   He does not belie the order, but he
fails to see that every order, in its visible aspects, is in
planetary contradiction to its central sun.   It is not indeed necessary that he should fail of this recognition ,
he has only to transcend the limitations of the partial
view, by which his consideration is confined to a study
of structure and function wholly with reference to environment, to see that the truths of this relation are the
disguises rather than the interpretations of life.   Morphology then becomes the science of creative transformations, wherein, as also in all functioning, it is not
the environment which determines life, but life which
makes its demand upon the environment.   The old
propositions will be maintained—expressing the visible
habit of inorganic and organic existence in terms the
most convenient and exact for the purposes of science
—but they will yield to their opposites, confessing the
truth of which they in their rigid outlines are denials ,
having indeed that reaction which belongs to life itself,
whereby all apparently fixed and inflexible certitudes
and stable embodiments dissolve into the unseen and
indefinable mystery from which they sprang.   All matter, in all its forms, has this solvency and release.

XXIX

The cosmic desire and expectation from the beginning reaches forward to incarnation.   This in itself
is an intimation of some special glory con-
Incarnation. summated in the flesh—the last and most
exquisite product of terrestrial culture.   Whatever of

descent there may seem to have been from the ethereal
estate of nebulous flame to that of the mute insensate
crust of the earth, we cannot but regard the progression of cell-life as an ascension, as if from the cinders
of extinguished fires some new flame had arisen more
nearly imaging the flame of the Spirit, since it had
breath, and in many ways witnessing that Spirit as no
star could do, nor the mightiest motion of the wind or
sea. This flame, which breathes in the vegetable as one
breathes in sleep, and which even there is aspirant, many-
colored and fragrant, and a flame of increase, in the animal awakes, and besides exhibiting a greater variety of
color and more wonderful fertility than in plant-life, has
will and sensibility. In animate life what marvellous
ascension—from the worm to the insect, from the creeping reptile to the hot-blooded bird which encloses,
possesses, and commands the element upon which it
depends, more buoyant than that which supports it,
seeming to be an embodiment at once of flame and air,
expressing heaven and echoing the heaven-song! The
animal seems to have won a kind of independence of
the earth, a show of separateness emphasised by its
power of voluntary motion. Its complex organism is a
deeper involvement than is apparent in less advanced
forms, and yet it seems to be the most perfect visible
revelation of the essential quality of life, as if in its
breathing and pulsation, in its spontaneity of motion
and feeling, and in its expansion and inhibition, it were
the living representation of that primordial manifestation which science strives to apprehend in its study of
the original constitution of matter. Since it is the visible realisation of the cosmic desire, therefore desire as

manifested in its activities and impulses naturally
seems to us the very image of the divine yearning in
creation from the beginning. So, regarding the most
perfect fleshly embodiment, we speak of it as "the
human form divine"; having reached the finest network of imprisonment, we seem at the same time to
have reached a critical moment of emancipation. as if
in man—the extreme complication of finitude and the
most fallible of all creatures, considered simply as an
animal and without regard to his peculiar psychical development—life for the first time assumed an erect
position and a divine gait. Thus always men have
imagined the divine after the human pattern ; it is an
inevitable idealism, and if it be the greatest of illusions,
it is one luminous with all the light there is for us in
the present order of things.

Nevertheless it is not an illusion in which the human
spirit finds rest; and though we can imagine no more
glorious forms for heavenly inhabitants. and St. John
in his Apocalypse admitted even the lower animals as
participants in the celestial ritual, yet is there the feeling of spiritual revulsion from the flesh as from the
world itself, so strongly expressed by St. Paul in his
use of the term carnal and in his assertion that "flesh
and blood cannot inherit the kingdom of heaven."
The finest cosmic texture which we know—the most
beautiful garment we see God by that issues from the
loom of time—is turned from as if it were also the
grossest. In it is stored all the sweetness of earthly
existence, a warmth and influence more magical than
is intimated in the forces disclosed in the chemist's
laboratory ; yet is it a glory that must pass, and in no

dissolution is there corruption more repellent, not even in the miasma of vegetable decomposition. But it is repented of before its divestiture in that new involution of life — that psychical synthesis which is distinctive of human destiny.

XXX

Following the line of thought thus far taken, we may not regard the human species as evolved from any other; and it is conceded by some of the most eminent evolutionists that there is not the slightest evidence of such a derivation nor any ground for its hypothetical postulation.

Life has no beginning or end, save as it is always beginning and always ending. Man, before his appearance as a distinct species in the specialisation of cell life, was not excluded from the series of transformations looking forward to his incarnation. In all specialisations he was a distinct species, his royal line of kinship being, like Distinctive Human Specialisation. that of Melchisedec, without beginning or end of days. That which he is now, in comparative physiology, is typical of his relative position from the beginning in all cosmic manifestation—a position which we can no more represent to ourselves in any definite conception than we can forecast what it will be in any future existence.

Man was not first an animal and afterward man. In the earliest stages of his development his animality suffered a kind of indignity from the psychical characteristics which ultimately were to give him supremacy, so that among animals he was at a disadvantage, lacking somewhat of that infallible knowledge which belonged to their instinct, and appearing less competent physically than many other species for the conflict with external conditions. A rational intelligence, such as distinguishes the man of to-day, transferred to that period, would have regarded the human species as ignominiously defective, and at a fatal disadvantage even as compared with the apes, from some variety of which he is thought to have descended; every conspicuous difference from these, including his want of a tail, would have seemed to emphasise his inferiority. To such an intelligence the law of the survival of the fittest would have seemed to put the human weakling *hors de combat.* Thus impossible is it logically to anticipate the creative transformations of life!

In the case presented, the transformation had already been effected, though its glorious issues were hidden beneath the masque of apparently hopeless weakness and ineptitude.

The human infant in gestation is seen to resemble, at various stages, animals of inferior species, as if recapitulating its own association with the progressive specialisation of animal life from the protozoan upward; but as we know that the infant is, at every one of these stages, human, proceeding toward a distinctive destiny heralded for it from its germination, so it is not unreasonable to presume that the progression thus represented was itself charged with the same distinctive destiny. Man as a protozoan was man, distinguished from all other protozoans, having that likeness to them

which the human germ has to the germs of all other
animals, one of appearance only.

We have been considering the illusions arising from
specialisation, from the progressive involution of life,
and increasing with the complexity of organisation; but
the ever more and more manifold veiling of life, certainly in the organic kingdom, is for us a progressive
revelation, while the visible appearance of the simplest
forms of existence is of all appearances the most de-
lusive, a blind masque, insinuating identity and sterile
unity, and confounding all diverse destinies.

XXXI

Humanity is in its specialisation inseparable from
the specialisation of Will and Reason. We here touch
the pivotal point of a new world. All divergent rays
are here concentrated and reflected; and it is thus
that the human incarnation becomes the
express image of God. From the long night   The Fall.
of time emerges the Logos become flesh, whose desire for incarnation has dominated the cosmic procession, making the universe the complement of himself. All other embodiment was the adumbration and
expectation of his appearance. How long he was
withheld as the special nursling of Elohim, or with
what fiery baptism he was tempered in that brooding
infancy which we call Eden, we know not. We know
him only from the moment of his flight from Paradise,
when began for him the cycle of wandering which had
been foretokened in the movement of all worlds. Per-

chance, if he might have turned and fallen upon the flaming sword, there might have remained for him forever the level world of innocence and simplicity ; but as easily might the Earth have repudiated her planetary destiny and have fallen into the sun.

That which we call the fall of man was in all primitive legends represented as his levitation rather, or aspiration, his entrance upon his proper destiny, and was associated directly with the development of his rational or discursive intelligence. He partook of the fruit of the tree of knowledge — that knowledge which distinguishes between good and evil. The story is one that shifts its shape and incidents and meaning according to the human mood. In the Promethean legend it is not the fall but the betterment of man that is intimated. The Titan (belonging to the Earth dynasty, which is in alliance with the human race against the jealous Olympian gods) steals fire from the hostile heaven for the benefit of man, who is thus enabled to start upon his career of progress. In the Hebrew legend there is a hint of Titanic help in the advice of the serpent and a suggestion of jealous alarm on the part of the Elohim who send in haste the cherubim to guard the tree of Life. The first use of clothing is also indicated, as the beginning of man's larger investiture. But the time when the legend took its final shape was evidently one of reaction against the artificial conditions of civilised life—one of weariness and dissatisfaction with "all the labor of man under the sun." At such a time man would seem to have lost some higher estate through vain curiosity and overweening pride, and to have eaten fruit, for his own sake wisely forbid-

den, when he surrendered instinct for errant and fallible reason and safe simplicity for the innumerable perils
of a haughty venture.

XXXII

But it was his destiny, and the very essence of it
was its psychical character. To all other animals
choice could have no rational meaning, since in their
selection the alternative is instinctively rejected. All animal consciousness is doubtless in kind the same as the human, and
there is in it an adumbration of reason, having, however, no properly rational field or career, as in the
case of man. But man, as at the same time mastering
all other animality, and repenting himself of his own,
has a psychical nature wholly unique. The lion, his
embodiment having been perfected, has no field of operation outside of his bodily functions. The corporeal
perfection of a man, on the other hand, is an utter
blank, from which no positive suggestion can be derived as to his peculiar terrestrial destiny; as blank as
was the Earth in her merely structural perfection as to
any suggestion of the flora and fauna of which she was
to become mother and nurse.

One of the most interesting studies in natural science
is the consideration of the transformation which vegetable and animal life have wrought in the earth : as in
the restoration by bacteria to the soil of elements
drawn from it and converted into animal tissue; in the
culture of the soil by earth-worms; in the erosion of
stones by lichens; in the storage of sunbeams by vege-

*A Singular
Psychical
Destiny.*

tables in coal deposits ; and in the building up of continents by lowly creatures living in shells, whose work is completed by coral germs.* But the terrestrial transformation wrought by man is much more remarkable, because it is effected through an arbitrary selection and adjustment, which, though in some ways fortunately inapplicable, is, in others, almost limitless. He cannot rival the earth-worm's ploughing, but he can make a garden of the desert, and reduce to temperate order the riotous wilderness. More rapidly than the lichens he reduces the rocks to dust. His destruction of certain species of animals and his domestication and improvement of others ; his artificial modification of plants and fruits ; and his diversion of watercourses, have changed the outward appearance of the globe. His temples and pyramids, his cities and towns and hamlets upon the land and his fleets upon the sea have humanised every landscape ; and even the mischief resulting from his wasteful destruction of forests and the blotches he has made upon the bright face of Nature are evidences of his masterful power to impress his mark upon the world. These are but the visible signs of his psychical supremacy—such as would be disclosed to the casual regard, and do not begin to tell the story of this new universe of man and mind. A visitor from some other planet who had no experience of a similar development would find in these obvious phenomena no adequate indication of their own meaning ; and a close scrutiny of details, disclosing temples,

* See *The Study of Animal Life*, by J. Arthur Thompson, pp. 21-26.

the edifices for varied social uses, the industrial machinery, the libraries and art galleries, the equipment of
museums and scientific laboratories, the insignia of political and military functions, the properties of diverse
amusements, and the paraphernalia of domestic economies, would bring into the view of such a stranger a
system of symbols requiring the most elaborate in-
struction for their comprehension, which would be the
revelation not only of what man has done for the earth,
but also of the uses he has made of matter and force
for purely human ends. A still closer study, even if it
were confined to the single department of literature,
would lay open the vista of human history and reveal
the marvellous imaginations and speculations of individual poets and philosophers, showing man as the
thinker and interpreter as well as the doer. In this
view the human, or what is the same thing, the psychical, destiny transcends all earthly contacts and material
uses rising to the concerns of an invisible world, to solicitudes and aspirations which overleap the physical
limitations of existence.

Like a celestial firmament above the earthly is this
new realm of Thought, whose tension is broken in the
precipitate of speech—the Word from the beginning
ultimately expressed in the articulate word. The rhythmic harmony of the animate, the incarnate, ascends
into overtones of psychical harmony. Here is a new
involution, a fresh embodiment—an adumbration, at
least, of what St. Paul calls a "spiritual body." The
tension here is a mystical unfathomable storage of potential energy, next immediately to the quick deaths of
the brain, but for it less directly all the world dies; it

is an ascension for which all the waves of cosmic life forever rise and fall.

The desire which has shaped and informed macrocosm and microcosm, ever sphering itself anew, and entering upon new tropes in its action and reaction, passing from order to order, each wonderfully diversified and co-ordinated, becomes now the ensphered rational Will. Every successive stage of the progression up to this point had involved additional suspense, more complex limitation, increased temperament, until in the starved deeps of ocean and upon the barren crust of earth the cell appeared. Extreme limitation, compensation, balanced resistances, gave organic life its opportunity; and in the development of this life that balance became more conspicuous, the physiological functions of the more complex organisms having a dualistic or divided action, as in respiration and circulation, and the interaction between the vegetable and animal kingdom maintaining a contrapuntal harmony. It was ever a more delicate poise of equilibration until, in psychical action, it became a deliberate volition in the subtle temperament of consciousness. What range of suspense from that of a planet like Saturn, which in the poet's fancy

"Sleeps on his luminous ring,"

to that of the spirit's contemplation ! Earth has her summer when she is at her greatest distance from the sun, latency and ascension being greatest when the patent energy is least, or when it is most in poise. So man, upon creation's outermost rim, has his psychical

ascension, his will, though under the extremest limitation, being the express image of the divine.

In the contradiction between man's position, as the
most helpless and fallible of all creatures, and his
destiny as the son of God, we confront the human
comedy, wherein the emphasis of time has its intensest
exaggeration, and the eternal familiarity its deepest
meaning.

XXXIII

In the human world the outer worldliness is repeated and outdone, having an infinite projection.
The multiplicity and variety of the physical universe
sink into insignificance beside this new series of involvements and complications. "The Father
worketh hitherto," and now man works, build-   The Con-
ing his superstructure above the divine foun-   scious Veil-
                                                  ing and Dis-
dation. Has God hidden Himself behind the   cernment.
veils of His world? Man has multiplied these veils,
whereby he has also hidden from himself his own essential self. He comes into a world of hidden fires
and broken lights, a world of interrupted currents and
of apparent stabilities, rigid to the point of frangibility ;
and this broken world he still further breaks ; his mind
is a prism, and what to his vision is already partial
becomes more discrete in his analysis, and most articulate in his speech.

In the specialised consciousness nothing begins
save by interruption or termination. Definition is by
boundaries, by the lines of cleavage in the brokenness
of things in time and space, so that judgment is dis-

cernment. We would have no definite conception of
light and no name for it but for its interruption, or of
any current save as it is broken. Whatever elements
there may be in the universe, about us or within us,
that are not thus discurrent cannot enter into the dis-
course of our reason. Dissociation seems primary,
and our association is of the dissociated elements, co-
ordination being the reflex of radiant diversification.
It is true that our first sensibility seems to hold all
things in a kind of confusion, but the progress of
intelligent perception is through discrimination and
comparison.

XXXIV

The psychical, like every other order, is planetary.
Not only are all other systems therein reflected and
recognized, but it is itself a distinctively human system
of thought and volition thrown off and dissociated from
the solar man, showing in the earliest period of its
development that fluidity and instability
which characterises the primitive planet, and
then gradually hiding its fires, losing its
clairvoyant transparency in opacity, and shaping its
firmaments. We have at last the superficial planetary
man, seeming to himself to have a motion wholly his
own; and this illusion is fortified by the fact that in
his sky (which thus differs from all other planetary
skies) the central sun is never seen. There is no
blindness, no opacity, like that of the extremely specialised planetary consciousness, wherein knowledge
becomes wholly relative, objective, partial, and limited

*The Plane-
tary Man.*

to the visible course of things—to the closed circuits
of physical and mental phenomena. In its extreme
rationalism it excludes the miracle and becomes entangled in the meshes of its own web, vainly attempting the solution of problems which are of its own
making, since they arise only within the network of
relation, association, and causation, whereby, as by the
links of an endless chain, it is imprisoned. The
strangest feature of this illusion is that the confinement is known as liberation; and such it truly is—
the planetary liberty of arbitrary selection, of choice.
Here, too, is maintained the likeness of the psychical
to the physical planetary system, in that the order
seems to deny its central principle: instinct is hidden
under arbitrary determination, the Son becomes the
Pupil, experimenting on his own account and learning
only by failure; the fountain is lost in the stream, and
essential attributes are disguised in the outward and
structural integrity.

"God hath so set the world in the heart of man,"
saith the Preacher, "that man knoweth not what He
hath been doing from the beginning even unto the
end."

XXXV

Behold how the illusions thicken and multiply in
this world which includes the phenomena of conscious
will and intelligence. Life, in these outer
courts of its temple, seems to deny its essen- Psychical
Illusions.
tial attributes. In itself spontaneous, direct,
immediate, it becomes the opposite of all these in a

secondary nature, where action and knowledge seem arbitrary, where they are relative, through means toward ends, all operation proceeding by indirection. Consider the contradiction involved in the necessity of making acquaintance in this casual and indirect way, as in a game of hide-and-seek, with beings we have always known. In our relations to other existence, what incongruity: that we should depend upon it for sustenance; that we should enter into alliance with it for our protection and into apparent conflict with it for very standing room! Life only is potent: whence, then, this guise of helplessness, this stress of concern as to means of life, as to provision for safety against impending perils? What strange mansion is this, against whose portals beat eagerly for entrance all human souls; and of those finding entrance how questionable their tenure! From the open sea what winds and currents drive against the reefs and rocks of a coast that is at once hospitable and forbidding, inviting to the shelter of secure havens, drawing also to shallows and shipwreck, the merest triviality dividing safety from destruction! And this human drift, which is the latest, with what reckless violence does it fling itself against the indurations of time, seeking a foothold, where with patient endurance it fortifies its position, cheerfully trying conclusions with things in this rude field of experimentation and adventure!

The greatest of all illusions dominating the mind of man in the world of appearances is that of his outward selfhood, eclipsing his inmost and essential personality. It is a selfhood which seems to him a complete estate, which he calls the Ego. He ignores the will and intelligence which have fashioned and informed his members, becoming at last sensibility and volition incarnate ; he ignores these as if they were not properly his own, and calls his only the mind he has made, and the will which he has formed and which he calls his character—just as he calls his only those corporeal motions which arise from his conscious volition.

What we thus term illusions are but the habits wherewith we clothe ourselves, the masques and varied costumes which we wear in the Comedy—the veils of the transformation-scene. What is within ? What is that Fire which never flames but is the ground of all flame ? What is that Light which is unbroken and knows no shadow ? What is that which itself flows not but is in the fountain that by which the fountain rises and falls ? What is that which is not born and never dies but is the principle of nascence and destruction ? We know not so as to name, and yet it is really all we know, the ground of all our knowledge. It can be stalled in no predicament. The Pantheist, Monist, and Dualist utter their names and definitions in the face of the Unutterable. To say that beneath all that is disclosed in our consciousness is the One Will and Intelligence—the indivisible soul of the Universe —is an assertion derived from our conception of a finite individuality. In the very essence of Life is that which gives the meaning to our terms One and Many, but not to the one apart from the other. Any predication which is not the absolute negation of all predicament brings us back into the outer courts of the temple — into our everchanging habit and habitation—into the pulsation of embodiment. We are clothed upon not with immor-

tality but with mortality; habit itself, whether of the
flesh or of the spirit, being, like memory, the resurgence
of a falling wave.   As we have said before, our old
Nurse from the beginning is both Lethe and Levana.

Man, more than any other creature, is by his desire
and his destiny (which are one) thrust into exile,
thrown upon his own venture, absorbed in his voluntary endeavor.   His is not the blind preoccupation of
instinct, but a wakeful, solicitous intention, engaging
every faculty of his complex nature.   For a time in the
infancy of the race he leans to the earth in a natural
piety and humility, worshipping Demeter, and looking
for help to the benignant powers of darkness.   But
how quickly his old nurse shows herself as Levana
rather than Lethe!   From the first, indeed, the urgency
of his peculiar destiny is apparent, driving him into the
far-country, and he stands face to face with his limitations—peculiar limitations upon which only human life
enters, and which are at once the source of his weakness and his strength.   By his very individuation he
is lost, and seems like one disinherited and at odds
with a rude, alien, and resistant world that tempts and
bewilders him.   Reduced to a state of pupilage he
must strive for all he would have or know, only those
doors opening to him at which he knocks.   From his
sensible contacts with the world he builds up mind and
experience, faltering into his intelligence.   His walking
is a series of falls, and he stumbles into all his progression.   Ignorance and fallibility seem to be the very
ground of his curiosity and aspiration.   Disturbance
becomes stimulation, resistance the measure of his
strength; that which is in the way becomes the way.

These are the very conditions of that destiny which begins in revulsion from animal instinct, a revolt involving shame and humiliation and defeat, a sense, also, of conflict with Nature — with her life and her death — but from these conditions arise the glory of the human world.

The solar man—the centre of this planetary psychical system—though hidden, is still the potential energy vitalising and illuminating the specialised individual will and reason and the collective social order which is the result of human effort and intelligence. This latent energy shines with native light through the rude dawn of social culture ; an informing divination and inspiration ; the initiation of mystical rites, with choral song and dance ; the spring of buoyant adventure and heroism ; the tender inward grace of faltering beginnings ; the plenitude of faith, making up for inexpertness and lack of outward vantage. Pessimism lies at the end of things, waiting upon facility, as the sense of vanity attends accomplishment.

As blind feeling, hidden beneath the specialisation of sensibility in vision and hearing, remains the living ground of the beautiful perspective developed, so that native divination which is buried beneath the constructions of the human understanding remains the living ground of the vast and varied rational perspective, being indeed the invisible and latent power which lifts man into a realm whose interests range in ever-widening circles from the hearth-stone to the remotest star. But it is the hiding of this power that accentuates the human perspective and makes possible certain peculiar conceits in the human consciousness — such as have been already instanced as illusions, all emphasising the apparent independence of that outward integrity which is built up by the individual and collective will, and which, as a whole, constitutes what we call the moral order.

CHAPTER II

THE MORAL ORDER

I

THE term *moral* is derived from the Latin word *mos*, meaning custom, and *ethical*, from a Greek word (ἔθος) having a similar meaning. Primarily these terms suggest wont, inclination that has become habitual, a spontaneous disposition. This spontaneity is apparent in the beginnings of a social order and in the first stages of æsthetic development. Human actions, like the operations of Nature, seem to fall into order of themselves, and with reference to some unseen centre of harmony. Choice, instead of being an arbitrary action of the will, is rather a dilection, accordant to the invisible harmony, a natural selection in the subjective sense of the term, a divine motion and passion, having also natural inhibition or restraint, corresponding to the modulation and temperament of the cosmic order. Nothing in the human world is vitalised save by the divine action and passion, and the vitality is not an endowment, it is genetic. In this view the problem concerning Free-will could not occur. We do not question whether the flower turns to the sun or the sun turns the flower, when these are seen not as two motions but as parts of one.

II

In the complex specialisation of the moral order this spontaneity is more and more hidden and apparently contradicted in the prominence given to arbi-

The Outward Standard. trary selection. In the Latin word *mos* there is the suggestion of *measure* (from the old root *ma*), so that one comes to say of his habitual conduct that it is not only his wont, but his rule; and in the social evolution the individual comes under a rule not his own, to which both his inclination and his reason may be subjected. The tendency is to substitute for flexible principle the inflexible rule. As the individual artificer works with plummet and level and rule and according to some rational plan, so does society collectively seem to build up its institutions in conformity to some outward standard and according to reason. Whatever is necessary to maintain the tone, health, and vigor of an organisation (a family, a tribe, a nation, or a confederacy of nations) pertains to its *morale* and determines moral obligations, and these obligations will be rigorous as against whatever tends to disintegration. Life will be more and more hidden for the gain of structural strength, until in the most complex of civilisations it will seem to be buried under its mechanical framework. In a merely superficial view the entire moral order seems to depend upon arbitrary selection, to be the result of experimentation, the sum of which we call experience. If we were confined to this view, absolute pessimism would be the only goal of our philosophy. Considering the moral order merely for what it outwardly seems to be, as summed up in man's accomplishment and what he aims to accomplish, our vision of the human prodigal would end in utter nakedness and inanition, just as any theory of the cosmic order confined to the study of structure and function would lead us finally to the view of an intensely cold and sterile space filled with dead worlds.

III

Seen only as shut into the field of his exile—of his conscious plans and efforts—the weakness of man is conspicuous, and his shame and misery intolerable ; no joys within these limitations can balance his pains ; the last word of all his speech must be Vanity! Elsewhere in the boundless universe there is no such sense of humiliation, as elsewhere there is no such capacity for vacillation, misadventure, and defeat. All other cosmic operation has its prodigality of a divided living, its error and defect, its vagrancy and avoidance of perfection, its swerving from straight lines and from regularity of form, but its procedure, however mediate, is sure, without indirection, never mistaking its course ; and this is true also of human corporeal and psychical action not under the control of conscious volition. Life outside of the field of arbitrary choice knows no outward rule or standard ; its order is of sure ordinance, a spontaneous co-ordination, involving no experiments, no misfits, and never missing the happiness of harmonious adaptation—constituting a world of everlasting loyalty

Peculiar Aspects of Human Experience.

and fidelity to the brooding Spirit, which is at every point unfalteringly wise and potent. Here there is, properly speaking, no experience, no cumulative knowledge. There is, indeed, variability of habit, discontinuity, transformation—a series of surprises delighting even an all-wise expectation; but, whatever the habit or change of habit, there is thereby no formation of a divine character or increase of a divine knowledge. The Spirit of Life *becomes* the universe, which is always and everywhere a fitness as well as a becoming, and as much so at the first (if there were a first) as it can ever be.

Human experience, on the other hand, is quite distinct from these cosmic phenomena, and, considering its scope and its aims, is widely different from that of all other animals. But the distinction is not so absolute in reality as it is in appearance. There seems to be a great chasm between a voluntary effort which in a brief period accomplishes its purpose and those physical operations which for the attainment of a similar end would take millions of years. What a vast period of time is occupied in the development of a human hand or eye! But the machines invented by man in a single century give him a hundred hands, and enlarge a thousand-fold his scope of vision. In making his engines and his telescope, however, he is compelled to avail himself of the properties of things, of natural forces and laws, and it has taken ages for him to learn these uses. He can arbitrarily regulate the speed of his machinery and the power of his telescope, but his water-wheel is of no advantage apart from the gravitation of water, and his telescope is useless apart from his living

eye. He cannot impart life to the products of his artifice, and such modifications of living things as are the
result of his deliberate adjustments are really brought
about by vital processes that in themselves are independent of his will. What he has gained has been acquired by slow and faltering processes, and as the result of innumerable failures. If we regard only the
ends which he consciously proposes to himself in his
experimentation, these in themselves are utterly vain,
having no value save in the living principle from which
all human aspiration springs, and which reaches its
true and living issue only as it overleaps his goals, disclosing their unreality—the emptiness of all which we
call a conclusion and accomplishment.

IV

This living principle is hidden—it is the secret disposition of our divine-human destiny, and when, in
some luminous moment, it shines through all its veils,
or when, in some flaming moment of transformation, the
vesture is consumed, then indeed our conscious plans and propositions are disclosed
as mere broken fragments, the partial seg-
ments of a cycle which is completed in a movement
that escapes observation. This hidden life it is—our
own very inmost life—which flanks every strongest for-
tress we can build.

Disposition
and
Proposition.

Now, when our old Nurse Lethe becomes to us
Levana, putting us away from herself, setting us upon
our feet and turning our faces toward outward goals,

she still attends us, though unseen. She lodges courage in our hearts to meet the bewilderments of a world
that at once tempts and betrays; she assists our faltering steps, making inert, resistant matter our support and our very fears an inspiration, deepening
our hearts through solicitudes, enlarging our strength
through travail. She it is who, though taking us often
back into the merciful oblivion of sleep, yet draws
about our busy day-dream a curtain, hiding from our
specialised vision both the fountains and the issues of
our life, and shutting us into our game of Hide-and-
seek, to which she gives zest by wonderful surprises,
showing us at moments of defeat gifts more precious
than those we have sought and lost. Following our
blush of shame because of something marred or missed
is seen upon her countenance a special grace—a favour-
ing glimpse of the Ideal. She sets us at our looms,
and though we weave but shreds and shrouds, text-
ures whereby we are clothed upon with mortality, she
sometimes gives us glimpses of the other side of the
web, where it is mystically seen as whole, or at least
suggests some beautiful inward integrity marvellously
contrasting with the apparent outside raggedness. The
emphasis of Time would paralyse our hearts but for her
quickening of prophetic hope, showing escape where we
behold only a barrier, and reserving as the largest of all
her favours that last release, when she sets our feet
toward the door of our dwelling, which they re-enter not.

While, therefore, our experience seems to us experimentation for the most part, so that we have come
to look upon the present existence as a period of probation and even to think of eternity as dependent

upon time, imagining some everlasting mansion patterned and determined by our shaping of its mere
scaffolding; while we magnify our exploitation and our
conscious manipulation of things, laying supreme stress
upon arbitrary choice and upon human responsibility,
yet such a view interposes an irrational chasm between
human existence and the general course of things. It
is especially a modern view, confirmed by the impressions derived from an extremely specialised order—industrial, political, and ethical—where artifice seems to
displace creation and the thing made the thing that
grows; where formal and lifeless mechanism is most
conspicuous, and where the ends proposed appear to
be as far removed as possible from such as lie in the
line of natural selection. The aggregation of people
in large cities, the accumulation of wealth, the artificial
conditions of civilisation, the absorption of so many
human lives in efforts to secure simple subsistence, the
magnitude of enterprises undertaken, the mastery over
natural forces, the devitalisation of industry through
the extreme division of labour, the secularisation of institutions, the tendency toward a social collectivism in
which the individual and the family shall be subjected
to a general control, and the supreme confidence of
society in systems of reform, and in the power of statu-
tory legislation—all these indicate the predominance
of arbitrary over natural selection. Free human will
and human responsibility are transferred from the cir-
cumference of a specialised order, where properly they
belong, to its very centre ; they seem to overshadow
all other factors of progress, and in the moral order
they claim the entire domain.

V

In the absolute sense there is no purely arbitrary selection, and what we call free will is so in appearance only and by virtue of limitation, being the ultimate specialisation of spontaneous Will, just as our reason is the ultimate specialisation of spontaneous intelligence. Choice seems arbitrary because of our consciousness of its most delicate poise and balance in a world of librations only less specialised; because of its extreme variableness beyond that of any other functioning in the organic realm, becoming sometimes even caprice; because also of its fallibility, which is associated with the empirical or experimental.

No Absolutely Arbitrary Selection.

Human experience is not divorced from human destiny, but is rather its masque, that which man proposes to himself, in the line of his phenomenal progression, disguising his secret disposition. All appearance disguises Reality. There is the Real Will with a hidden purpose deeper than any particular intention—a Real Reason deeper than is shown in any definite rational process. The Cosmos hides the true Logos—that light which lighteth every man, and which is the Light of the World. In man as in the world it is the genetic or creative that is hidden.

Nothing falls outside of this genetic reality, though everything thus seems to fall outside of it in our conscious representation of a world to ourselves. There is no power or knowledge separate from it. We say that we make something common by communicating it;

in reality it can be communicated only because it is common. We call that general which is the result of our generalisation, but the ground of generalisation is the genetic. There is no familiarity out of the family or home. What we see as a divided living is genetically the abounding life—the ground of multiplication; what seems to us under restraint, in tension within walls visible or invisible, is genetically the bounding as it is the abounding—the ground of all inhibition; what we see as form is here formative, informing, and transforming, and that which we know as order is here undistributed harmony. Herein is the eternal life — to know the Father and the Son—eternal kinship, eternal familiarity. Life in this transcendency needs no chart or guide or standard; has no prudence or economy or any moral virtue; cares not for any structure or type; it stands for all that falls as for all that rises, for evil as for good, for divestiture and destruction as for embodiment and growth, for mourning and fasting as for a festival, for the old as for the new, seeing both as one —the reaction in the action, repentance in regeneration.

VI

The genetic eternal life is the ground of all action and reaction, which are proper thereto in a sense wholly indefinable in our specialised con- sciousness. No predication we make con- Genetic Authority. cerning the action and reaction as seen in the visible world—a world of suspense, where beginning and end are regarded as separate—is applicable to the invisible genesis where death and birth (if these terms could there be used) are inseparable. Neverthe-less the genetic and eternal meanings dominate and give significance to all phenomenal existence—consti-tuting the bond of kinship which makes the whole uni-verse the Father's House, whatever our illusions of flight and exile, of freedom and integrity.

The genetic is revealed even in its veilings, and in the illusions of our divided living it has a varied and beautiful disclosure—a confession in its very denial. In the inorganic world the hiding seems more complete because we see that world only on its dying side, in its descent and diminution for the ascent of organisms; though even here we see death as genetic, the barren becoming fruitful, the desert inflorescent. In the invo-lutions of the organic the revelation is clearer and more intimate, the abounding creative life showing itself in ascension, growth, and procreation, in all forms of in-crease signifying its original authority.

In the physiology of advanced organisms this genetic authority is conspicuous, though its source is hidden, residing in a system of cells quite distinct from those of the general system, and having a sacred inviolability and immunity, secure from waste and expenditure in ordinary functioning; an empire far withdrawn from the outer courts of the temple of life into its Holy of Holies. The nebulous and comparatively unspecialised germ plasma dominates the whole embodiment, being the source of its motion and passion—the physiological symbol of eternal life, of that which was from the be-ginning and which is to come. This tenseless potency is surrendered only in such germs as come into embodiment, wherein again it is hidden, since of its kingdom there is no end.

In the human world the dominion of this principle is supreme. We see it in the primitive worship of ancestors and in the symbolism of all sacred mysteries; it is associated with all human heroism and chivalry as with native virtue and piety, with the beautiful in art as with the sanctities of home—the one lien which Nature has upon man in the most artificial conditions of his civilisation. It is not of matter, but of the spirit, or, rather, it is of matter because it is of the spirit. Man is the incarnation of the spontaneous Logos, whereof all else in Nature is only a less specialised manifestation; and the essential idea of the Logos is genetic —it is Sonship.

We dwell upon this conception of the genetic, as the basis of all natural selection, of vital destination, of harmonious ordinance, as the Reality beneath all appearances of individuation and altruism, of separation, conflict, and association, because it impresses upon our minds and hearts the sense of a universal homely disposition of things; also because in any order, and especially in the moral, the Appearance is regarded rather than the Reality.

Birth itself is a break with the eternal, and the first deliverance of infantile consciousness, separating the me from the not-me, is the beginning of a perspective of wonderful beauty and variety, pulsing with the life it veils, at once an involution and an evolution, a folding away of the self and an unfolding of it, and in the same movement an assimilation and a differentiation. Individuation is by inclusion and at the same time by

exclusion. In the many-mansioned house when one
door is opened another is shut, when a fold of its curtains is drawn another side of the fold shows a withdrawing ; a falling at one point is a rising at the next,
so that the whole architecture is a succession of living
waves. It is a Passing Scene—a constant Trope—a
turning and a re-turning. Every point is the centre of
repulsion and attraction—one the complement of the
other, and both together making spheres of matter and
cycles of motion. In this flowing equation the very
contradiction of the two sides represents their identity
—but also their interchangeability. This Protean metabolism is the outward revelation of the essential consubstantiality of all things with each other and of all
with the Father. The hunger of every desire has its
satisfaction, partaking of what becomes its own, only
because that which is appropriated was already its portion—a part of itself—as God is the portion of every
creature. Each desire is in the line of its special accords, but the largest harmony encloses all these. The
divided living is a specialisation, but in the eternal
reality all that the Father hath belongs to the prodigal
as to the elder brother. The line of a family in genetic succession is one of special accords—of special attractions and repulsions—and in this line vital destination is pronounced and plainly seen to be inevitable ;
but in its destiny, and really the largest part thereof,
is included the dominion and service of the universal kinship which it seems to exclude as alien. The
more particular specialisation in the predilection of pri-
mogeniture, both as to honors and sacrifices, privi-
leges and responsibilities, is really an exaltation of the

genetic principle itself, whose glory is summed up in the First, Only, and Eternally Begotten.

The real Personality is a mystery transcending any possible mental analysis. The analyses attempted in recent psychological speculation have but one result— the multiplication of personalities within what is usually regarded as the embodiment of one ; and in the monad-ological theory the multiplication is extended indefi-nitely. Such notional analysis is itself a specialised rational process, hiding the real truth, stalling it in the numerical predication, and leading to just such irrecon-cilable contradictions as result from the consideration of a divisible Space or Time. The principle of divisi-bility is itself a genetic mystery, which is disguised in the mathematical process. We say " apart from," ex-pressing distance—in a real apprehension we would say " a part of." * Separation, if it be a vital depart-ure, is the breaking of a union which still remains one, including the fragment. This is not Pantheism, unless St. Paul was a pantheist, declaring that in God " we live and move and have our being." To think of our-selves as " without God," and of our wills as other than indissoluble from His will, is the falsehood.

Matter is not acted upon by other matter, as indi-cated in the statement of physical laws, or a spirit by other spirits: the action, including the reaction, is *in* each but *of* all. There is no dominion of quantity, no majority. The infinitesimal germ balances the universe, and, while in its individuation it seems to hermetically

---

* The word *part* has itself the genetic meaning : *pars*, from *pario*, allied to *portio*, which has the same root as the Greek ἐπόρον (*gave*), perfect πέπρωται (it is destined).

seal its integrity, it has infinite endosmosis—an openness to all currents; and though it seems to gather
where it has not strewn, yet its vital use and possession is the appropriation of its own : it is in debt for
no endowment, save as owing and owning are one.
There is no fund of potency and wisdom, or of life—
only an eternal fount of these which we call our Father,
to Whom our vital relation is not one of accountability.

Neither is the individual, as a living being, indebted
to Society. The bond is vital, conferring upon the
individual no rights and devolving no duties. Rights
and duties are not pertinent to natural laws, but only
to conventional regulations and adjustments growing
out of the specialisation of social functions. The son
has no duty to the parent from his sonship, but because of his tutelage. We do not associate debt and
credit or any merit with parental instinct or natural
piety. The bond is so close and intimate that these
terms are not adequate to its expression. But the
period of human infancy is prolonged far beyond the
limitations of such a state in other animals, and more
extended in the advanced than in the primitive stages
of progress, so that the family, though primarily a
natural institution, involves a care and culture over-
stepping the bounds of instinct, varying according to
circumstance and guided by rational motives. This
special tutelage assumes functions whose exercise has
an important bearing upon social interests outside of
the family; it is moral and educational, demanding
outward rules and standards and requiring obedience
and conformity.

VII

The patriarchate is the similitude of the Father's House, and in passing from it to the tribal organisation natural vitalism was still maintained. When, in more complex grouping and a more special- ised social life, a conventional bond took A Superficial Retrospect. the place of the primitive sacrament of kin- ship, human progress assumed new aspects; and in our retrospect of this departure it seems like an en- trance upon a new world. As in the first development of human intelligence the rational is differentiated from the instinctive, involving a peculiar weakness and also a peculiar strength, so in the beginning of conven- tional institutions this differentiation is more marked, and with the weakening of living bonds there is fortifi- cation of the social structure. History and the science of history deal mainly with this structure—with the deliberate human efforts engaged in its elaboration, and with the outward conditions affecting the growth and decay of social systems. While the thoughtful student regards this structure as a living organism, a superficial view discloses the mechanism only, which has indeed the semblance of an organism, but seems independent of the general course of things—a drama whose scenes are shifted arbitrarily by a human will that has somehow broken loose from the universal harmony —a by-play of Destiny rather than its ultimate expres- sion. We are apt to review the history of mankind in the lights and shadows affecting our conception of the present situation, beset by problems of every sort that

seem to defy solution — a transitional situation the
issue of which no man can see. Human progress,
thus regarded, appears to be from the vital to the unvital, from the strength of a flowing life to the brittleness of mechanical stability, a constantly greater surrender of potential energy for structural completeness.
The traces of that golden age of humanity, which our
imagination vaguely locates in some remote past, retreat before our longing backward vision until they are
lost to view.   We assume that at some unhappy epoch
in the very dawn of history man abandoned a first
estate of innocence and was himself abandoned, thrown
upon his own resources of will and reason, and compelled to win his way upon an earth accursed for his
sake, through harsh conflicts with a hostile nature and
with hostile aliens of his own race, and under the over-
whelming shadow of jealous gods whose angels fiercely
guarded his forfeit heritage, and who baffled his heaven-
piercing aspirations with such confusion as befell the
builders of the Tower of Babel—gods who were wor-
shipped because of fear and in a perpetual ritual of
propitiatory sacrifices.   We picture to ourselves this
Marplot of the universe, this Protagonist who by his
first arbitrary choice involved a world in death and
woe, as forever after shut into such edifices as his arbitrary choice might erect for his pleasure, protection,
and use, and in all his ways brought face to face with
the Death whom by denying he had invoked, and with
the dread monsters following in Death's train or anticipating his approach.   The development of this victim
of so many pursuers appears to us the result of his
antagonisms, and especially of his conflict with Death,

whose terrors become the chief inspiration of life, giving swiftness and suppleness to his flying limbs, sharp-
ness to his faculties, and cunning to his intelligence;
deepening his imagination; and prompting him to
build monuments that shall survive his brief existence. Even the procession of generations appears to
be a defiance to the arch-enemy, each one that passes
smiling in the face of the great Destroyer and pointing
to its successor.

Beholding man as thus the arbiter of his own destiny, scheming, ambitious, and selfish, in all his struggles seeking and slowly gaining vantage by the sheer
force of his own will guided by the light of a mind
built up by experience, and considering the solicitudes
and apprehensions attending his first rude exploitation
of a refractory world, wrecked in his own ruin, we follow with a feeling of mingled pity and admiration his
ruggedly adventurous career from his first attempts to
clothe his conscious nakedness until his habit has hardened into a mailed armour covering his infinite vulner-
ability. While all living things erect and expand their
structures in apparent defiance of gravitation, and he
likewise counteracts this force in his upright frame,
using it and breaking it in his gait, yet in his artificial
constructions, dealing with inert materials, he must
build with level and plummet and upon a firm foundation. In place of the sureness of instinct he must estab-
lish for himself the certitudes of reason, and in accordance with these adjust every detail of his individual
life and of his more elaborate social economies. In his
reason he must find compensations for its own fallibility, the rule for righting himself against his many falls.

As, according to this view of history, religion is born of fear and the love of mastery is nourished by incessant antagonism against hostile elements and forces, so the shyness, suspicion, and cunning arising from apprehension and developed in constant efforts for resistance and protection bring one tribe into war with another or several tribes into alliance against a common foe; so that it is through conflict, through conciliations to avoid conflict or to solidify attack or defence, and through treaties following the issues of conflict, that the larger social groupings are formed.

In this more complex organisation a new element of weakness calls for a new system of fortifications. In the single tribe the blood of kindred was the sole fountain of law, and morality was hardly distinguished from natural piety. Restraints were vital. We see from the scriptural account how God is said to have treated the first murderer, sending him forth as a wanderer, but setting upon him a seal for his protection—a very different procedure from that enjoined by the Mosaic law for the government of the aggregated Hebrew tribes. The arbitrarily devised statutes, for the regulation of peoples acknowledging no living bond of social obligation, seem to us to have been wholly arbitrary, and we represent to ourselves a deliberately wrought political system, with conventional allotment of property, of rights, and of duties, and even the fabrication of a secondary conscience. In a word, formal justice, regulating every social economy, takes the place of the natural, living control; and the substitute appears to us so inherently weak because of its conventional character that we inquire how it was reinforced. The weakness itself, the dire necessity, would have prompted to rigorous discipline, to a severe penal code. The very frailty of government would have enthroned the governor and hedged him about with divinity. The priest would have stood at his side and forged the thunderbolts of heaven for the enforcement of the civil edict. The military sacrament, displacing that of kinship, would have stood for protection not only against foreign invasion, but against internal revolt. The inflexible barriers separating castes would have given solidarity to the social structure. Empires would have grown by conquest, securing peace within their borders, and fostering the culture of art and science and jurisprudence. Thus Rome became the mistress of the world, nations seeking alliance with her even more for the benefits of her stable dominion than from fear of her victorious legions.

In this benignant atmosphere a sense of mastery succeeds to that of weakness, and the poet forecasts a new golden age of world-wide peace, stability, and equity. The will of man has conquered Fate, and has caused to grow in the garden of Experience fruits of virtue outrivalling any products of Nature's fairest fields. It has especially transcended Nature in bringing to bloom the thornless rose of Merit—a flower to which no instinct may lay claim and which may not fitly lie within even a mother's bosom—the mead alone of Virtue's brow. Nature can bring forth only new things; Man, by the exercise of arbitrary selection, makes a better world, a worthier manhood; against her vagrancy and defect he shows his moral rectitude and the faultless symmetry of his art; against her prodigality his prudence ; against her spontaneity and surprises his care and calculation; against her undiscriminating beneficence and pain his evenly measured equity—not yet fully realised, but perfect in its ideal aim and sure of ultimate outward completeness.

Has, then, the Promethean dream come true, despite the jealousy of Jove and the arrows of Apollo? And has the spark the old Titan stole from heaven grown into the soft flame of human amiability, courtesy, and easy tolerance, subduing ancient enmities, and newly limning the face of man into this frank mien that shows no traces of the ancient fear and furtive cunning? What betterment!—a term which Nature knows not in its moral sense — and all from Choice, the device of human will and reason in their revolt from a first nature and in their emancipation from its bondage ! The old gods have new faces—not only in their fashioning from the sculptor's chisel, but as feigned by human thought. Long ago, in the inspiration of its revolt against the nature-gods, the Hellenic mind had found a new goddess—Athene Parthenos—the unbegotten virgin, springing fully equipped from the brain of Zeus, having no taint of that injustice which runs in every line of Nature, and fitly representing the completeness of outward integrity—the Queen of the Air, the patroness of Athens and of the culture whose procedure is by arbitrary selection. Now there is a kindlier thought of all the gods. Perhaps they were benevolent, but not omnipotent, themselves limited by a relentless fate, which, like man, and possibly with the help of man, they could only slowly overcome ; perhaps they, too, were struggling against refractory mat-

ter for the establishment of justice and for the exclusion of darkness and death and evil from the universe.

But even while the poet dreams, the vast empire is
crumbling, soon to be broken into a hundred fragments. The Age of Gold again recedes into the irrecoverable past, and philosophy bewails the vanity of
all the labour of man under the sun.

A new civilisation begins the building of its temple
of Justice—an association involving new impulses and
motives which tend to the enlightenment and emancipation of all·peoples. But the leaven is hidden, and in
this new world, as in the old, there are cruel wars, feuds
of caste, the development of selfish interests and of
altruism as the expression of educated selfishness;
slaveries are abolished only to give place to others
harsher and less vital ; and, regarding the merely outward aspects of all human economies, we seem, at the
end of this nineteenth century, to be approaching an era
of sterility like that reached in the development of the
earth's structure before the appearance of cellular life.
From such a consideration, and in accordance with the
cosmic analogy, we might reasonably look for the advent of some entirely new order of terrestrial beings as
distinct from humanity as the organic kingdom is from
the inorganic.

VIII

This view of human history, while containing much
that is true, is partial—distorted by false dogma and
false philosophy.

It is true that fear has been an important element in the human drama, especially the fear of Death. But in the dawn of conscious endeavour, in the earliest intimations man gave of his peculiar destiny, this fear was not an oppression and did not beget panic; it was the shadow of a brightness. Sensibility trembles into its outward manifestation. The eye is at first dazed and troubled by the light to which it awakens. Yet the organised embodiment reaches in man its greatest eagerness and hunger. It is the urgency of vital destination rather than a deliberate choice, an inward boldness showing itself at first in outward shyness. As we have already seen, the advance of all organic existence is toward a greater peril, a more conspicuous mortality; and in man the venture trespasses all limits, inviting number-less risks. How violent must be the subjective uncon-scious (or sub-conscious) will which maintains its secret disposition despite the conscious avoidance, con-flict, and solicitude — characteristics that are, indeed, much more apparent in the attitude of modern man than in that of a primitive race! The difference be-tween a Roman of the time of Marcus Aurelius and the immediate offspring of the fabled wolf-nursed Romu-lus and Remus is as great as that between the dainty, comfort-loving kitten and its fierce feline prototype, the lion, in whose heart was lodged a native courage, generosity, and temperance, sharply contrasted with the cowardly alarm, the developed cunning, and treach-erous playfulness of its sleek descendant.

Native races show the mark of an urgent destiny, which is hidden more and more with the development

Fallacies In-volved in the Superficial Retrospect.

of consciousness, and they are not fairly represented
in the degenerate cave - dwellers, the easy preserva-
tion and exposure of whose bones, in their secure re-
treats, have misled or, at least, unduly impressed the
anthropologist. We are apt to overestimate the conscious weakness of men in those periods when conventional institutions first began to overshadow natural
control, just as we exaggerate the artificial character
of those institutions. The potential energy is at its
maximum in the least specialised stages of human
progress; and though the outward weakness, leading
to much faltering and stumbling, is manifest to our
historical judgment, we also discern indications of a
natural heroism and enthusiasm which gave buoyancy
to enterprise — a sublime confidence not to be accounted for save by reference to that vital destination
which defies external conditions, transcending experience. The human will, in its more spontaneous movement, had no help from a logical plan, but in reality
determined the plan itself, establishing that rhythm
which was essential to the social order, and which, in
its elaborate distribution, is modulated, losing the violent impulse in the regular pulsation.

The social evolution, primarily an involution, while
producing a world of its own, distinctive in all its
aspects, proceeds by natural selection as does all cosmic development—the selection in either case being
determined by the living will and not by environment,
which is indeed itself only the result of this spontaneous and harmonious determination. In the social
as in the cosmic order there is a progressive modulation of forces, and tendency to uniformity and ap-

parent stability, until in the extreme poise of the
human will we observe an apparent indifference, trivial
casualty, and easily shifting caprice, corresponding to
similar indications of fortuity and inconsequence in
happenings upon the surface of things in the material
world. It is here, in the field of the extremely trite
and partial, quite divorced from any manifest desire or
meaning, that we become casuists and fatalists, seeking for omens in what is least pre-calculable, making a
lottery in the chances of indeterminate allotment. The
original divination, however, was based upon the spontaneity rather than upon the mere fortuity of these
happenings, which, because of their dissociation from
any definite mental reckoning, were thought to betray
a hidden divine disposition. The chance at the surface was thus associated with destiny at the centre.
To the ancient mind the "fortuitous concourse of
atoms," as the initiation of a universe, would have suggested divinity. How often it happens that the Gate
of Accident opens upon some movement hitherto concealed from our conscious observation, but which has
been going on behind the curtain of the "common
light of day." When we touch chance, we broach
God.

Destiny is only another name for Life itself—Life
considered not as a fund upon which the will draws, but
as itself personal Will, and sufficient to its own issues,
as not only from eternity consenting to what in time
engages its forces for resistance and conflict, but from
eternity determining its embodiment, its limitations,
and death itself.

When, therefore, a critical point in human history is

reached, like that which separates civilisation from the simpler native conditions which preceded it, we need not regard the transition as abrupt and involving the sharp distinction given to it in our logical analysis. We look upon civilisation as a kind of second nature of humanity, but it is not the less nature, nor less a part of human destiny, being indeed that which outwardly distinguishes man from the brute creation. Neither is it less spontaneously determined, however the genetic quality may be hidden in artifice and contrivance. There is nothing in the dry tree that was not begun in the green—not even its dryness. At the extremity we see in fixed form what at the centre is formative in the genetic sense, and the dead leaves falling disclose the seed, so that genesis is proclaimed at the last as at the first.

If it could be supposed that the type of existence known to us as the human had failed of an earthly manifestation, no other type could, through whatever environment, have taken its place; and all that is distinctive to this type in its actual development was pertinent thereto from the beginning. The terrestrial headship assumed by man and his mastery, in Deed and in Interpretation, were intimated in his simplest estate. The conscious human Accord, in the full perspective of its harmony — to which no other note in the universe is alien — will sound true to its original key, whatever the variations or dissonances in its procession.

But for the upholding and sure efficiency of vital destination, life would be at a loss at every critical turn. Even the ant or bee or beaver, if there is a

break in its instinctive construction, has some flash
from the broken current which gives a guiding light,
helping it to a kind of conscious recovery.   In human
experience, by its very terms and limitations, an incessant discurrence calls for constant recovery—so that
the entire existence comes to seem a fault, demanding
redemption.   Every illusion of the phenomenal world
arises from this brokenness, inwardly made whole when
we see others as ourselves, aliens as kin ; this conscious
vision being possible only when the perspective is complete.   Destiny, the eternal life, has the mystical vision
of the kinship from the beginning even unto the end.
In the successive sphering of self, family, tribe, nation
—each individuation being in its special involvement a
kind of seclusion and dissociation, presenting the aspects
of conflict—there is the vital co-ordination of the planetary system of humanity, the distribution, throughout
the series, of the harmony which *becomes* the system.

Repulsion — the dissociation already alluded to as
necessary to integration, so that to the family the neigh-
bour seems an alien, and still more therefore the adja-
cent tribe—is shown in conflict ; but the social instinct
was always a re-agent in a corresponding attraction,
without which there could have been no conflict—that
is, none that could be distinguished from the predatory
and destructive warfare waged by the brute beasts for
the satisfaction of physical hunger.   Hospitality toward
the stranger was always stronger than the hostile disposition.   Isaac was the type and Ishmael the excep-
tion.   It is because man is more social than any other
animal that he is so pre-eminently a fighter; and his battles have even an element of romance in them not asso-

ciated with struggles for mere material advantage or
for the "survival of the fittest." In the lines of destiny
affiliation lies beyond as well as before the struggle,
and those who have been shedding each other's blood
mingle their blood afterward in a solemn pact, establishing kinship, which is to be still more closely cemented by
intermarriage; thus the civil intercourse that follows
has not wholly passed beyond the living bond. To the
victors belong not only the spoils, but the vanquished
themselves, so that, though a man may fail to be his
brother's keeper, his victim's he must be; and sometimes it happens that the situation is reversed, as when
Rome conquered Greece.

In the series of social integrations there is in each
some point of departure, of flight from its own restrict-
ed economy, toward something outside of itself. Desire
is itself altruistic. Reproduction, even by fission in the
lowest organisms, is the becoming another. This altruism is transformed into that of nutrition, wherein the
hunger of one individual seizes upon another for assimilation. Marriage is out of the family, often out of the
tribe, as in the Roman seizure of Sabine wives. "More
than my brothers are to me," is the expression of
friendship in all times. Thus life confesses the larger
kinship. All dilection is vital, and the rational element
involved is only its light, not its inward motive. It is
probably true that integral exclusiveness begets shyness and human contacts take first the outward appearance of antipathy, but it is the sympathy which is
inwardly dominant. The plunge into the cold stream
leads to an inward reaction in the vital current, and so
to greater warmth. It is because of the dominant in-

terest promoting harmony through discord that native
warfare developed generous sentiments and a discipline
afterward of great value in civil administration.   The
first slaves were captives taken in war, and their capt-
ure, being the alternative to their extermination, was
one of the alleviations of warfare.   Their subsequent
domestication " set them in families " and developed in
both masters and slaves the loyalty and affection natural to so intimate familiarity.

However complex the grouping, the family remained
in all its sanctities and tender emotions, and was an
important factor in all strifes and alliances.   In the
most complex and formal of all ancient civilisations,
that of Rome, we see in the last pages of its long record how persistent to the end was the worship of the
Lares and Penates and the care of the ancestral tomb
within the sacred precincts of the home.   During the
period of Roman civilisation scarcely a single animal
was added to the number of those which had been do-
mesticated in primitive times ; but these tamed beasts,
in so far as they were directly associated with the land,
and the land itself, could not be sold ; they were sacred
to the family.   In all ages one's country is his father-
land—*patria*—this term continuing the semblance of
the patriarchate, as all economy is, by its very etymol-
ogy, associated with the household.

The civil economy grew as naturally as the domestic,
and was from the first sustained by the urgency of
sentiments and interests which, transcending human experience, were its ground and not its product.   There
was no need of new reinforcement from any source
not already existing.

Religion cannot properly be said to be or to have been a necessary sanction of the moral order. Primarily morality was in no way distinct from religion. The secularisation of government, ethics, art, and philosophy went on *pari passu* with the progressive specialisation; and at the same time religious expression was in like manner specialised and in the same degree, itself in its outward form as much an apparent departure from and contradiction to the central spiritual principle of human life as was every other manifestation. It is only because of this departure that religion has seemed to be even incidentally a sanction of morality.

For the sake of clearness and at the risk of repetition, we must here revert to considerations already advanced in previous sections of this work. The original sacrament of kinship—the fountain of primitive piety, God-ward or man-ward—laid no more stress upon justice than does Nature, save that it was not, like Nature, impartial in its inequity. It claimed indulgence from the human or divine father rather than justice — excessive and exclusive indulgence. With the expansion of kinship the limits of exclusiveness were also widened, looking forward to the idea of the All-Father —a spiritual idea, the perfect realisation of which is the kingdom of heaven, whose inequities, whether of bliss or of pain, are as impartial as those of Nature—a kingdom, moreover, of living righteousness rather than of formal rectitude.

The illusions connected with the phenomenal world —*i. e.*, the world as represented in our consciousness, and as affecting our volitions directed toward outward ends—contradict, or seem to contradict, the Reality of

eternal life as apprehended by that consciousness and
determined by that will whereby we are the partakers
of this life. These illusions of the broken world to the
broken mind are inevitable, are vital. It is as if human destiny were itself thus broken and specialised,
making for us the beauty of color and sound and
speech and thought and feeling—at the same time also
the defect in these, that in them which, while essential
to all integration or limited embodiment, must work a
dissolution of every synthesis, wearing the bravest
vesture to rags.

These illusions belong to the very beginnings of
structural development in religion, art, and ethics; but
in these first aspirations, where the difficulty is the
greatest, and where the view is narrowest and there is
the least help from accumulated experience, the potential energy is miraculous, overleaping barriers, lifting
easily the burden of life, with a strength to spare that
transcends the task, and solicitudes are not oppressive.
The first sacrifices in religious rituals are not propitia-
tions but festivals; the first art was spontaneous, and
the pursuit of virtue easy and natural. In the golden
dawn the prodigal sets out upon his journey with no
grave misgivings. The sense of facility comes with
the descent, when the uplifting force of life which once
lightened and vitalised the whole structure is being
withdrawn, yielding it to gravitating and destructive
tendencies. In this dull twilight, full of solicitudes,
the illusions of time imprison and oppress. The bub-
bling fountain that became an impetuous torrent is
swallowed up in the dry sands of the desert.

Every manifestation of human life passes through

this cycle. The reaction is in every moment, hidden
at first but finally conspicuous. The structure gains
upon the life, absorbing more and more the conscious
will and attention, until it seems all in all; the impetuous current bears man on to the completeness of integral form, which he regards with pride and strives to
hold at its noontide culmination of beauty and strength,
its full content: this is the height of his illusions,
which now have wrapped him in their luminous veils,
becoming his whole expression, his very thought and
language; but while, in conscious complacence, he rejoices in his integrity, in his formed character, in his
complete art, in his argent-rounded thought, in his established polity, and in his consummate religious rite
and dogma, his inmost will repents itself of its accomplishment, and the reaction becomes outwardly evident
in induration and senescence.

The illusions, then, that arise in the human consciousness from the specialisation of existence and
of consciousness itself, pertain to the whole phenomenal world, including human experience; in their
beauty and glory they are associated with the passions
and aspirations, the attractions and repulsions, the as-
similations and jealousies that are involved in every
integration of individual and social life—fluent in the
superabundant vitalities engaged in crescent organisation, and fixed in the mature and stable structure,
where they are stereotyped in scripture and speech, in
established customs and codes, in the formal certitudes
of science, in the canons of arrested art and impulse,
and in the suspended inspiration and ritualistic expression of a settled faith; and in their graver hues

they blend with the purple shadows of dissolution, when stability itself is seen to be the flimsiest and raggedest of all the veils hiding the eternal Reality. That which we have been urgently persuaded to call something is brought to naught; no trace is left of outward goal and object—our very habitation and investment gone.

No form of life can claim pre-eminence over any other as escaping these illusions. No wise virgins of Religion can give of their oil to the foolish virgins of Art or Philosophy or Morality, where all alike are shut out from the Bridegroom's presence, save as in every room—in Academe or Hall of Judgment as in a Temple — He is given lodgment and met or overtaken on every path of the devious pilgrimages. To all alike the veils that hide are the only revelations ; and all alike deny as well as confess—Peter the same as Judas.

In the building up of any order the spiritual principle is veiled and apparently contradicted, whether the order be religious or moral; between these no distinction arises in human consciousness before each has been so specialised as to take the form of a definite system: and always the prevailing characteristics of the one are those of the other. What men at any time feel and believe socially is precisely what they feel and believe in their religious life, the rule of their conduct showing their thought of the divine. If they have a living righteousness, from hearts loving, forgiving, not judging, generous not according to exact measure, without servile fear of others or a desire to inspire such fear toward themselves, then to them God has this same living righteousness, from the same disposition

of heart. What men think it is right for them to do
they regard also as the righteousness of God. If they
are satisfied with formal justice and with conformity to
outward standards, then they deem such satisfaction
an essential feature of divine government.

It cannot, then, be properly said that religion is the
sanction of the moral order; it would indeed seem
more rational to derive religious doctrine from the exigencies of that order, since those features of the latter
which grow out of its peculiar limitations come to be
dogmatically associated with divine action in a sphere
where such limitations cannot be supposed to exist.
In reality religious practice and thought have the same
tendencies as all other practice and thought—the rite
and dogma becoming, in the specialisation of a system,
as formal and unvital as an outworn state ceremony or
a stale maxim of experience. It is not that the rite or
dogma are essentially lifeless or insignificant, but that
in their fixed form, their integral completeness, they
have confined their life and meaning within the form,
which has itself lost plasticity, and that as an expression of the human heart they have become automatic
through vain repetition,

"Like a song of little meaning though the words are strong."

A creed may express a universal truth, a spiritual reality
in itself so profound as to lie at the very root of life—
such a creed as is expressed in the simple phrase *Our
Father*, which, seen in its genetic reality, transcends
space, time, and causation, and can never be outworn.
But within what narrow limitations may this creed be

held? We need not go back to find its provincial limitation in tribal theology or the early Hebrew theocracy;
it is equally implied in the latest *Te Deum* sung in
all the churches of a civilised country because of a
great national victory. In every social organisation
less inclusive than that of a universal brotherhood this
simple creed must be denied, and in the competitions
of every practical economy it is irreparably broken
and compounded. It is urged by those who desire a
revision of our religious creeds that these should be
adapted to the advanced conditions of human progress; but it is by this very adaptation, which is a constant necessity in order to a *modus vivendi*, that their
essential principle is contradicted. While social organisation at every stage of its progress brings peoples
nearer together and expands the sentiment of human
brotherhood, developing a cosmopolitan sympathy, yet
it at the same time stimulates competition, multiplying
its opportunities in the ever-widening field of industrial
enterprise and commercial exchange.

We cannot here consider the possibilities anticipated
in the dreams of socialism, and which may indeed
transcend those dreams when the sentiment of human brotherhood becomes universally prevalent. We
are here confined to a view of social economies as
they have been and are now organised; and in
this view it is evident that both the religious and the
moral sentiment accommodate themselves to the conditions of social organisation, though in so doing they
contradict themselves, slay the prophets, and crucify
the Lord. It may more truly be said that they become
that organisation, in all its exclusions and inclusions—

its strifes and affiliations; they are genetic in their operation, and in becoming that which contradicts themselves they only express the tropical action and reaction proper to life itself.

Thus human experience, which, in a superficial historical retrospect, seems to depend so much upon arbitrary selection, following some rational plan consciously devised, appears upon a closer study to be as spontaneous as nature, having its roots in the quick ground of a life invisible and inexplicable. Its possibilities are incalculable, and it is as difficult to trace a logical plan in its past as to forecast its future. There is no science of history, and our philosophy of humanity as of the individual man is confined to a study of growth and decay. Our mental analysis and our imaginative constructions fall short of the hidden purpose, which is shown, and is yet to be shown, only in the issues of Life itself—Life creative, genetic, transcending causation. What we see at any period of history, in so far as we truly see anything, is some portion of humanity in the stress of social integration, all its vital forces engaged in the process, eagerly, passionately, and, with feverish excess of zeal, violently seizing upon all earthly materials and boldly annexing to the terrestrial realm the celestial and infernal; or we behold it in the relaxation of these energies, in a process of decline or degeneration. The types differ — as the Hebrew, Greek, and Roman in the ancient world—and with them the kind and degree of accomplishment and the character of dissolution. The same external conditions affect different races in different ways, and an

extensive movement, like that of the mediæval crusades, involving many peoples, produces certain results
in some countries, and quite diverse or even contrary
results in others. The genius of a race or of some
individual leader standing for a race — a Cæsar, a
Mohammed, or a Napoleon—determines the political
complexion of a continent, upsetting all previous calcu-
lations. Our host we have to reckon with is not Logic
but Life—a vital destination determining distinctive
types, temperaments, aspirations, and jealousies.

IX

The categorical imperative—what we call conscience
— proceeds from the practical or living Will and
Reason, and the form of the mandate is plastic,
according to the vital determination. Its
variation is not from one fixed proposition
to another, thus presenting itself as an incongruous
series — it is a variation in the disposition of life.
This imperative is a bond in integration, and in death
an absolution. Take, for example, the family relation—of husbands to wives and children to parents ;
this involves obligations which are vital to social integration, and which are varied in passing from a
patriarchate to a more complex society. He who pronounced against divorce, when asked whose wife she
should be in the resurrection who had had seven successive husbands in this life, regarded the question as
not pertinent to the state awaiting us which should
know no marriage. He taught that the commandment

Conscience.

to observe the Sabbath was for man and not man for
the commandment—a truth applicable to all commandment, which must be a vital requirement.

All selection is for a living use in the most complex as in the simplest social order. What in a nomadic habit is a quick taking and leaving becomes in
more stable communities a long holding and a slow
release ; the suspense is emphasised. The co-ordination of an elaborate system affects the sentiment re-
lating to property, reputation, rights, and duties. The
categorical imperative reaches out to every manifold
detail; and in all relations honour yields honesty and
faith fidelity. The fruits in the garden of experience
are growths and not mere fashions arbitrarily wrought
by cunning artifice ; even the flower of Merit has its
living root, however much in its nice human culture it
may have lost of the wild flavour of its native stock.
The honey of the hive is, not far away, the wild honey
of the tree. The grape in the autumn sunshine seems
to invite the bruising of the contrived human press that
so, by its ultimate fermentation, it may yield its finer
spirit. The things men try for or by which they are
tried are in themselves nothings, nor has the trial itself
any meaning apart from the spontaneous life which is
the ground of all experience. The doors we knock at
with importunity, or which we unlock by the mechanic
leverage of our keys, open to the treasures of life,
which have no wealth save for that life's native abundance. Opportunity and temptation have only the subjective significance given them by the heart's desire.
Our existence, in so far as it has worth and beauty
and dignity, is made up of passions, which, however

modulated in temperament, must for their freshness
be forever renewed from their inmost source, and
which are never very far removed from the native attractions and repulsions that originally determined
their spheres and orbits. We do not prize unim-
passioned goodness. Culture is worthless save for
its secret inspiration.

Accordingly we find that human sentiment, in the
most refined civilisation and brought into its most
orderly realm, is not so much a revolt from
Nature as in many of its moral aspects it
seems to be. Conscious restraint, or rational
control, regarded as a moral merit, is but a specialised
form of that inhibition which, unconscious and un-
trained, is yet a more potent and surer bond in all
natural operation. There is no such temperance attainable as that which Nature has spontaneously—no positive purity like that of passion itself. The conscious
voluntary effort in this direction has its ground in the
inward temper.

Convention not a Revolt from Nature.

X

Rectitude rigidly conceived, whose sign is a straight
line, is not a living ideal, but in every real motion it
is a notional standard which is shunned as
well as sought. Righteousness has its outward notional standard of formal justice, but no real
righteousness is ever truly represented by the even
balance of the scales. The flowing equation of life
suggests compensation, but cannot even for an infinitesimal moment rest therein. There is no motion

Formal Justice.

but for some preponderance that disturbs equilibrium. A single inflexibility in any order would destroy it. Justice even in its own field refuses to be just.

Having reference to illusory appearance, we think that our aim is to secure rectitude, justice, stability; but, as in nature there is no point of rest, so in human nature satisfaction seeks emptiness as eagerly as emptiness satisfaction. Men neither desire to render or to receive absolute justice, having therefor a contempt as for anything Laodicean. Even in business, a dollar is parted with for the sake of or received at the risk of usury; and the zest of all commercial exchange is the thought of vantage on either side; and as a benefit is given as well as taken, the barter resembles that beneficent and complementary interchange always going on in Nature. Any withdrawal from this commerce, by a refusal to expend or to produce, checks the natural increase and tends to sterility. The general disposition of the merchant is toward an overflowing measure rather than the close, hard bargain. Men love to act the part of the host, and are gracious enough also to cheerfully receive hospitality. In such amenities they console themselves for the necessary restraints upon their generosity imposed by the conditions of trade; and one of the sweetest graces of home is that there one may give and take with no thought of return. Few are they who keep within bounds even in the performance of duties, especially of those duties which involve sentiment; few who are careful, prudent, or thrifty enough to manage a business for themselves, or who, in subordinate positions, do not overdo service; few who are as conscious of their merits as they are self-

reproachful for their faults; few indeed who do not
carry altruism to a mischievous extreme, regarding the
affairs of others more than their own.

As every natural tendency is, as Emerson says, "over-
loaded," so in human conduct every sentiment engaged
is overcharged and runs into excess.   Men are enthu-
siastic and intemperate in their patriotism; they will
not measure their loyalty. They retain, as long as they
may, devotion to a personal sovereign for the vitality it
seems to have as compared with the service of even an
ideal commonwealth. They cling to an intimate household economy, even if it involves slavery, or to a feudal
bond, until in the relentless course of progress they are
perforce emancipated. In a democracy based upon universal suffrage the masses of men will yield to the mastery of leaders rather than to that of ideas, regardless
of material interests at stake, and will fight for a prejudice sooner and with more zeal than for an ethical principle. They will, indeed, more readily follow the leader-
ship of good men than that of demagogues, if they are
thus brought into an association of sympathetic fellowship instead of being invited by argument. Humour
and prejudice are more vital than logic.

In religion also the human disposition rebels against
a measured service, and the desire of the heart could not
be satisfied by a divine ministration exactly compensa-
tory. Man wants nothing of divine justice; his appeal
is to a partial and special providence, to paternal indulgence. If punishments apprehended are incommensurate with the offence, in his imagination of them, the
hoped-for rewards are equally incommensurate with any
possible merit. In all genuine faith from the begin-

ning. Grace has been the essential divine quality—the basis of a forgiveness as free as the fallibility of man is inevitable.

XI

Progress in all systems has tendencies that seem to contradict human sympathy and faith. Every human synthesis, become a sphere, hardens at the surface, and the superficial contacts in the field of outward experience disclose the hardness and intractability and attrition. It is here, in a field of effort—training, culture, severe discipline—that the sense of arbitrary volition is intensified; here, in this scant and rocky soil, that man cultivates the hardy virtues which are prized exceedingly as the fruitage of patient toil; here where he stumbles most that he idealises rectitude; here where he is a pupil, gaining knowledge and power by slow acquirement, that the aim of all life seems to be improvement, betterment. Here are his varied and exquisite pleasures as well as his pains; his successes as well as his failures; the flush of pride as well as the blush of shame.

Limitation and Induration.

The induration, like the limitation, is a Mercy, the express favour of a life lost for an exquisite sensibility and capacity through which it is consciously recovered. This human incarnation—the latest and most wondrous of all creative miracles known to us—surely it was the divine longing from the beginning, gained only after many avatars. Eagerly the water became the wine and the wine the blood, until in psychical man the universe is reflected as in a microcosm. He stands upon an

earth dulled and stilled for his sake, stands where the
sun meets the dark glebe and gives forth a warmth not
known to interstellar spaces ; he rejoices in the same
intimate warmth through all his pulses and in the
breath of the tempered and tempering atmosphere.
Light is broken for his eye and sound for his ear, and
the whole world for his varied hunger. After the tre-
mendous clashings of the elements and in the midst of
clashings still continued, but which he perceives not,
there is this armistice for his peace, this suspense for
the happiness of his dwelling. The isolation of his individuality is a blissful seclusion into whose penumbra
only the predestined guest may enter, who, however repellent in his first guise, is afterward surely unmasqued
as an accordant friend. The day is meted to his
measured effort and the night to the measured rhythm
of his sleep. As his pulse repeats itself for his body's
growth, so does the pulse of memory and habit for the
gradual increase of his experience. The successive
days, like the successive moments, are

> " Linked each to each by natural piety,"

and he does not see what wondrous change is in the
transition, and that what seems to him continuous is a
series of deaths and resurrections ; the dead quietly
buries its dead, and each day is new, not overmuch
troubled by the ghost of yesterday or the shadow of
to-morrow. The blessed oblivion of the past and ignorance of the future secure the clearness of the present,
giving to each moment its particularity and that sufficiency which it properly has, since in it is eternity, as
in every particular is the universe. In this comforta-

ble seclusion he does not hear the grass grow and is not sensible of the swift motion of the earth in space; his communication and correspondence are well guarded, so that but little of the joys and sorrows of the wide world enter to confuse his individual portion, which is itself, whether sad or joyous, an allotment by littles and tempered to his limitations. As pet names take the form of diminutives, so our intense delights and sympathies are inseparably associated with our limitations, with what is petty and partial in our lot. Fidelity in small things is the test of the faithful, who, though they may be made rulers over many things, still hold the small things nearest and dearest, finding in close intimacies the homeliness of existence. The wife of one's bosom, the few friends of choice—these for nearly all men make up the sum of all that gives joy and worth and dignity to the earthly life, and the virtues and duties born of these are christened again with colder names for larger associations. Here is the nucleus of all social order, preserved fresh and tender by the very hardnesses of elaborate system, as the soft children are guarded by the toil-hardened hands of parents, whose wearisome routine encircles them with a wall of defence. All indurations are walls about the free play of life within. So fortitude becomes sacrifice. The more complex and formal and unyielding the social order is in its outward structure, the more nearly does it secure the inviolability of the individual and domestic seclusion. The sign of life within the veil of the temple seems reversed in the outer courts, becoming the contradictory sign. The flexible, the flowing, the spontaneous becomes there the fixed, the arbitrary, the inflexible. Grace

there becomes Justice, the Trope of life relentless
Atropos. Thus life is fully clothed upon with mortality.

As it is the complex vertebrate animal which is also
the warm-blooded and delicately nerved, and as the
tough shell encases the sweet kernel or vital germ, so
that ancient civilisation in which the family institution
was regarded the most precious and important, and in
which close intimacies had the deepest sincerity, so
that it gave to the modern world the name for piety
and every other virtue, developed also from the sacred
domestic penetralia the most complete system of public
functions and laws, the highest dignity and most inviolable obligations of citizenship, the most binding
soldier sacrament, and the toughest fibre of an imperial
structure, thus becoming the very backbone of the
world it dominated.

But the hard envelope about the seed must be
broken for the seed's germination and new abundance,
contributing in its dissolution to the sustenance of
the fresh growth, as in its outward completeness it
served for protection ; so the induration of all human
systems is the indication of their maturity, their readiness for death ; their suns at apogee have proclaimed
a new summer. The systems, like generations, pass
away, not because of their imperfections, but rather because they have reached such perfectness as their scope
has permitted ; not to give place to the better, but to
the new. In this passing, that which seemed stable
and inflexible becomes the flowing ; that which seemed
complete discloses its corruptibility ; all that has been
formed or acquired, whatever its excellence, beauty,

and loveliness, is brought to naught, save for its service of descent—its liberation of the spirit.

> "So God fulfils himself in many ways
> Lest one *good* custom should corrupt the world."

XII

In the completion of the cycle is confessed its spiritual principle, which in the stress of structural formation and functioning was obscured and apparently contradicted. *In articulo mortis* is uttered the true countersign to the eternal verity; and Death Confesses Life. it is seen that stability, fixity, immutability, and inflexibility are not pertinent to the eternal life— that these are terms which destroy themselves by natural termination and recourse.

To the eye of sense, regarding the worth of the structure as belonging to the edifice itself, seeing beauty and truth only in the formed thing—the formed mind, the formed character, the acquired experience— and the good of anything only in its possession, this vastation of a system seems utter vanity; but to the spiritual apprehension the loss is wholly gain—redemption, rehabilitation, a new creation; the eternal life, itself the truth, beauty, and charm of all that is visible, depends not upon any structure or acquisition.

It is because the eternal life is in the bright day that we ask for its continuance and regret its decline; also, it is because of this eternal life in it that it cannot stay; but that life is in the darkness as in the light. Happily our conservatism, sound and wholesome as it seems, nay, as it is, at the noontide of mature integrity, like the fixed fibre of the strong oak, is itself the habit of induration, ready to fall into the routine of descent and release, and serving as it falls. So turns the planet into new dark, new dawn. So turns the Wheel of Life.

It is only in our conscious representation to ourselves of life seen in a partial arc of its cycle that even in ascending movements there seems to be a conflict with death. It is here that death is included in its essential meaning for the constant renewal of youth. It is a part of our planetary opacity and confinement, and because our attention is fixed upon outward uses, that we regard evil as merely disciplinary and our present existence as peculiarly a probation—the contracted ante-chamber of eternity. The emphasis of Time prevents our seeing that our existence now is as truly grounded in the eternal as it ever can be, and that this is the ground of our reconcilement with all that we contend with and resist.

XIII

If our exile were real, if we could really leave the Father's house, if by some chasm Time were divorced from Eternity, and if human existence were wholly experimentation, consciously regulated and in *The Eternal in Time.* its entirety determined by arbitrary choice on a rational plan—as from partial aspects it seems to be—then indeed might we pray for absolute annihilation. In this view the moral order would be a system of inextricable confusion. If we can believe in such separation of humanity from its Lord that our life is hidden elsewhere than in him, then is inevitable that other belief, formulated in the extreme rationalistic specialisation of dogma, that there are dread realms of unutterable woe forever excluded from the divine presence and from the operation of divine laws and uses. If the material is separated by an impassable chasm from the spiritual, then may we accept the dualism of the Manichæist or adopt the scepticism of the biologist who asserts that matter only is eternal and that the entire realm of life is but a fleeting moment of cosmic time, a shuddering pulsation that for an instant disturbs the monstrous and heartless mechanism, an alien dream as inexplicable as it is transient. If his rectitude, his formed character—that outward integrity which he builds up for himself—is at its very best man's only blessedness, then is his experience vain ; if that whereof he is ashamed or that of which he is proud, if what he consciously shuns or what he consciously seeks be the full measure of his evil or of his good, then, in the superficial jaggedness of the things wherein he is entangled, is his destiny the most trivial of inconsequences, the ultimate caprice.

Not thus is he to be accounted for, and never in the depths of his spiritual being has he thus accounted for himself—as if he were a fragment of the world, appearing suddenly upon the ocean of existence, moved this way and that by varying winds and currents and by the whims of his own variable and near-sighted intelligence, and then as suddenly submerged beneath the

waves. He never had a spiritual philosophy which did
not make him one with the Eternal—which did not
make him the measure and explanation of the world
rather than the world the measure and explanation of
him—one in which the scope of his evil and of his
good did not embrace all evil and all good. In him
alone did life awake and think and speak, but not thus
did he forego his share in the eternal silence. Whatever his forfeit, it compromised the universe, and engaged all the powers of the universe for his redemption. No transaction could in its scope be too far-
reaching to be commensurate with his eternal interests.

XIV

The moral order must be referred to a spiritual
source, and whatever its contrary aspects, those are
such as characterise any order when seen in the light of

Psychical
Progress de-
pends upon
Spiritual
Growth.

its central principle. Regarded as a whole,
*the moral order is that cycle of human experi-
ence which, beginning in a flesh-and-blood kinship, is completed in a kinship which embraces
the universe.* Whatever it may seem to be in any part
of the cycle, it must in its totality be the outward expression of man's spiritual destiny. Conducted to its
completeness by any rational plan or by rules derived
from experience, it would be as remote from the Kingdom of Heaven as is the embodiment of Confucian
ethics in the Chinese social system; but if we conceive
the psychical progress of man to include his spiritual
growth in that garden of which the Father of Spirits is

the husbandman, and to be in its largest expression a
harmony whose centre is in the regenerate heart of a
divine humanity, then must this progress as a whole
transcend as well as include those constructions of
human will and reason which lie within the limitations
of experience—must indeed so far transcend these as
in its regeneration to be a repentance thereof, a repentance beyond all the natural repentances in the
series of creative transformations, bringing in the new
heavens and the new earth wherein dwelleth the living
righteousness.

These considerations which, in so far as they are
based upon a Christian philosophy, more properly belong to a subsequent chapter, are here introduced as a
protest against the identification of human regeneration with any possible outward accomplishment or integral completeness. But it is natural and consistent
with all analogies to regard man's psychical progress
as, in its mightiest reaction, associated with his redemption.

XV

Regarding the moral order as grounded in a spiritual
principle, we see the working of this principle in what
seems most arbitrary and conventional. Our plans
and charts of life are not merely subject to
revision, but they become parts of a dissolv- *The Hidden Life.*
ing view, the material world itself becoming
spiritually solvent. We cannot but fix an intent and
expectant gaze upon the objects of our striving; but
even while we look there is a change like that which

comes in dreams, and some hidden hope is answered
that is often contrary to the conscious expectation and
always different. There is a real world nearer and
more intimate than that which lies next to our eyes or
our hands. It is as if we were projecting our opposites, really yielding when we seem to resist, and releasing what we seem to seize, some deeper dilection contradicting the apparent choice and taking the evil
which we outwardly reject; so that, while we are load-
ing the scapegoat with our sins to bear them away into
the wilderness, there is something within us that takes
sin itself into that ancient confessional, wherein it is
conjoined with all dark mysteries and finds its reconcilement with the eternal life. What outwardly seems
weakness and shame is inwardly glorified, becoming a
part of the creative transformation whereby the quickening spirit moves to issues registered, indeed, in time,
but known for what they really are and mean only in
the council-chamber of the Eternal, where the Son is
one with the Father.

CHAPTER III

ASCENT AND DESCENT OF LIFE

I

" Remember now thy Creator * in the days of thy youth," says the Preacher, recognising the nearness of youth to its mystical source, as if in the ascendent movement of the fountain it might feel its motion ere it moved—as if through a gate not yet closed it had some vision of unwasted brightness and power.

Birth and Death.

Wordsworth associated childhood with intimations of immortality, though, as presented in his sublime ode, these intimations are those of an Eternal Life rather than of immortality—the native sense of that life as an unseen ocean whose waves are heard beating upon the shores of Time, " though inland far we be."

In the scientific view birth is most intimately associated with death. Thus, in the series of creative specialisations, sex appears simultaneously with death. Reproduction is a katabolic, or descending, process, the matrix is a tomb, from which Childhood is the resurrection. The highest organisms show most complex dying as well as most complex living; and in

---

* In the Hebrew the word signifies " Well."

every physiological operation the dying lies next the living process; thus the metabolism goes on, nutrition turning and falling into secretion and secretion stimulating nutrition.

Looking upon birth as the beginning of an organism, or the apparent beginning, and upon death as its apparent conclusion, then the whole term or cycle of its visible existence is the interval between these; but the extremes which we thus separate in thought are in every living moment of the organism brought together. The most significant fact disclosed by recent embryological research is the intimate connection of death with birth. Death permits birth.

In the most complex forms of life both death and birth are specialised and accentuated and are, moreover, prolonged and elaborate periods. In certain species, between the lowest and highest, the death of the parent seems to be the immediate sequel of the parental function, thus conspicuously emphasising the katabolic or mortal characteristic of the reproductive process.

II

The earth, before it could be the dwelling-place of man (of man as we know him) had come into a state of suspense and temperament, wherein her veiled potencies had a novel and varied manifestation—
Specific Preparation for Ascent of Organisms. the cosmic habit and constitution of a planet, showing what would almost seem a new kind of matter. As the arrest of her flight, bringing her into her destined orbit, disclosed grav-

ity, so the hiding of her heat permitted the molar attraction to enter into a free play of molecular affinities
hitherto latent in the primal expansion and tension.
The heat radiated from the contracting sphere is specific ; indeed, all the energies manifest in this ultimate
constitution of matter have a microcosmic specialty,
and in their planetary transmutation are released for
new tensions, diversely, multiformly, and minutely expressing the old theme in temperate and discrete ar-
ticulation.   It is as if what Plato, thinking of the generic, meant in his conception of Ideas had become
species, diversified in bewildering variety, especially in
organic existence.   The very distance of the planet
from the sun seems to permit this free play of life
upon its surface, as the departure of heat from water
permits crystallisation, or as the arrest of nutrition
brings fruit and seed.

In this complex hierarchy of Nature discrete accords
are sustained, so that they fall not into indifference
and confusion ; degrees of excellence are marked—of
truth, beauty, and goodness ; individual sequestration
and tranquillity are secured, and for each life a way—
its own that no other can take, and yet open to accordant intimacies and correspondences ; and in the
psychical involvement life acquires a feeling of itself
and a conscious control, the liberty of its dwelling.
Everything becomes special—birth, existence, death,
providence itself.   Space and Time are but the room
allowed for the play of action and interaction within
an appreciable scope, and the varied seasons through
which all things pass in their limited cycles.   For all
living things repose, like work, is special, giving to

night and sleep and oblivion—to all kinds of release
—peculiar and grateful meanings.

Moreover, in this ultimate constitution of matter, we
note a special latency of forces, a *vis inertiæ* of the elements—terrestrial insignia of the primal potency. Elements which readily combine at ordinary temperatures
are set far apart. Thus it is said that there is iron
enough hidden within the earth to wholly deprive the
atmosphere of its oxygen if it were all exposed at the
surface. Receptacles are provided for storage and
immunity, and walls for protection. In living organisms vitality includes the physical and chemical processes, holding them in suspense for its own ascent or
allowing their disclosure in its descent, just as the expansive power which we call the centrifugal force in
the solar system includes and veils gravitation until
the limit of expansion is reached, when the reaction is
disclosed.

At the last point of descent in the specialisation of
inorganic matter, which we call dead—at the point of
barrenness, appears the plasma of ascending organisms.
Science for the explication of undulations or waves of
energy (the forces themselves which we call heat, electricity, etc., being diversified according to wave-lengths)
postulates the ether as a vibratory medium pervading
all matter—the atoms of matter being vortical motions
of this ether. These vortical motions are likened to
smoke-wreaths, versions which are at the same time
introversions or retroversions. We would prefer to say
that life pervades the universe, and to designate these
motions of the ether as the tropic action and reaction
proper to life itself—an evolution which is at the same

time involution.   For whatever we may interpose between the phenomenal world and its Creator—whether
it be the absolute of abstract metaphysics or the ether
of physical science—the medium itself demands explication, and we must in the end confront the mystery of
the Eternal Life.   In the organic plasma, then—in this
nucleus of a universe which, seen by us as ascending
(since we are a part of the ascension), we are willing to
call living—there is the action and reaction of life.
We may think of these as internal motions, representing them through such images as to our limited understanding seem most adequate for their expression ; but
the only real apprehension of them we can ever have
is through their own expression in living manifestation.

If we consider this protoplasm as a material substance having certain properties and certain chemical
constituents *plus* vitality ; if we think of it as an involute in which all forms of vegetable and animal life
are held latently and implicitly, awaiting the stimulus
of environment for their evolution and taking such diverse shapes and functions as may be determined by
mechanical and chemical resistances and pressures, we
reckon without our host.   For this substance in its
apparent homogeneity and indifference does not more
completely obscure its possible issues than it veils the
unseen spirit unto which it is plastic.   It is because of
its apparent simplicity, insignificance, and characterless-
ness that it is susceptible to the infinite potency of the
abounding life which is to become the finite fulness
and variety—of that same abounding life which gives
the ether its pulsation.

This protoplasm, as already intimated, lies next an

utter barrenness in the inorganic world—next the win-
ter-like stillness and calm severity of natural elements
and forces hushed and checked for some singular Nativity.   All the travail and prodigal expenditure of
Nature, all her gracious descents await here in peace-
ful silence to catch the whispered longings of a new
Desire, which shall call for further and more special
ministrations.

III

We have noted that in organic specialisation there is
a physiological insphering, incavations for the reception into the organism itself of this descending ministration of Nature, hungry receptacles which
are at once tombs of decay and matrices of
life—the dying being thus intimately brought
next to the living.   The life of the organism
demands the sacrifice; it not only includes death, ever
multiplying and deepening its capacity therefor, but
gives it a physiological character as distinguished from
merely chemical disintegration, so that the descent
conforms to the physiological ascent which it promotes.
Moreover, each part of the organism thus nourished
suffers disintegration for its own functioning, and even
during a certain period gains structural strength through
this intimacy with death.   The waves of psychical ascent in like manner rise next to the quick deaths of
the brain.   Capacity and involution are for ascent;
faculty, function, all evolution, for descent.   The complete physiological term for each organism is a cycle the
curvature of which is determined by the limit of capacity.

*Organic In-
volution: a
Capacity
for Death.*

IV

When Euripides said that what we call living is
really dying, he expressed a truth as scientific as it
is poetic. What is it that we especially call living?
Is it not our complex functioning, our development in-
dividually and socially, something associ-
ated with our waking hours rather than with Our Living
a Dying.
those of sleep, with the expenditure of en-
ergies rather than with their expansion and absorption,
with the exercise of trained faculties, with active heroism and passionate romance, with the contests in the
arena, rather than with the crudeness of infancy, the
dependency of pupilage, the inly-folded dreams of
youth that give no outward sign, and the mimic conflicts of the gymnasium and palæstra? But wakefulness
is mortal exhaustion—functioning is a release of tension, like that of a watch which serves as time-keeper
only when it is "running down." No work is done
save by bodies that fall or in some way give up potential energy; all development or unfolding—what we
call evolution—is a descent. We know the tree by its
fruits, but inflorescence and fruition, beginning from
arrested nutrition, belong to the falling life, to its diminution.

Expression, all definite and visible manifestation, is
a witnessing—a martyrdom. Evaporation becomes invisible, but we see the descending rain, the flowing
stream, the crystalline ice: contraction, solidification,
and fixity of structure show the degrees of falling. In
the material world these processes of descent are most

in evidence; and their precipitate—what is known to us as dead matter—is an ever-present object of vision and touch, inert, resistant. Therefore it is that gravitation, which is the physical symbol of death, seems to us the prime universal force, and weight is made the measure of value to the disparagement of levity, being associated also with importance and impressiveness. The root of the Hebrew word for glory signifies heaviness. Solidity and stability bear down upon us with a like emphasis of force and pressure, becoming also the basis of confidence and firm support. So Death which draws us down becomes a prop against descent, the means of protection and fortification against his ruinous assault.

Mechanical work, like the functioning of living organisms, is made to depend upon this "dying fall." We accumulate gravitation by damming a stream, barring and accumulating its gravity, and then permit its operation as a driving force. We confine the tension of steam and then release it, regulating the escapement to suit our purpose. We even imitate the organic stimulation of nutrition through waste, as in the electric dynamo the reinforcement of the tension is increased by the larger outlet in expenditure.

The record of human history is a Book of Martyrs; the vista is lined with ruins. The beginnings of all races are lost to view. The biographer of eminent men searches in vain for traces of the child that is "father to the man." Our own first years are hidden in oblivion. The fountain of youth eludes discovery, escaping even contemporaneous observation. Our present is known to us only as it passes. Thought, regarded as a definite manifestation, is a precipitate; and the formed mental structure, like the formed moral character, is a mortal framework that needs support against its tendency to fall, as does any material edifice; indeed the time inevitably comes for its maturity, induration, fragility, and ruin — a season of autumn, postponed only, like the decay of the body, by nutrition and the stimulus of expenditure.

Death is, then, so inseparable from life that we speak of one in terms of the other; and in an external and objective view we must think of all action in an embodiment as finally taking upon itself the appearance of decrepitude and diminution—the original potency lost in impotence. In this view every embodiment is a prison-house, gradually closing in upon life with an absurd conclusion.

V

In reality, the involution is the tension and confinement, and development the graduated release. An invisible reaction in the ascending movement determines the limitation of every organism after its type—that is to say, in its special accord Limit from Reaction. as part of the cosmic harmony. It is a bond in the expansion and fixes the bounds apparent in the development; it controls the method and defines the shape; it establishes the curvature of the cycle completed in descending processes of evolution. Thus reaction seems to dominate action—hidden in ascent and conspicuous in descent. The smaller cycles of activity illustrate this domination of reaction as does

the full term including them. Thus every expert vocal-
ist knows that inspiration controls expiration. The elasticity is in the inbreathing, the withdrawal, the rebound,
the undoing (as in sleep), the moment of the heart's repose—always the vanishing side, as into creative void.
Invisibly, or subjectively, the limitation is the seal and
commission of power, but objectively, as seen in structure, it is a barrier — the sign of that impotence into
which it descends; and it is in this outward view that
the confinement oppresses or harasses and seems like
an entanglement full of hard knots that make us quer-
ulous and beget in us miserable solicitudes. Here it
is, and associated with the sense of imprisonment, that
problems arise to vex our souls, concerning life itself—
that life which transcends the prison and yet so seems
to belie and contradict itself within the narrow hedge-
ment. Thus our queries about a future life take their
very form and color from our cloistral structure—like
those which the Sadducees, who believed not in the
Resurrection, put to our Lord. We are apt, like the
Sadducees, to ignore the peculiar conditions of our
confinement, and most of all the fact that it is spontaneously determined by Life itself—by our own inmost and essential life, which is one with the Logos
from the beginning. Our distorted views of the present as well as of a future life arise from this ignorance,
which, as in the case of the Sadducees, is radically a
lack of faith in Life's proper reaction—its resurgence,
since the reaction is an ascent completed in descent,
a flight completed in return, a repulsion finally disclosed as attraction.

Concentrating our attention upon the visible world,

upon human development regarded externally, in relation to its environment, we are lost. Life's own insistence upon its limitation is so sure that it controls every form of thought and action. The closure is effective. The limitation is special, as if holding us to the key-note of a particular harmony. The protoplasmic basis of organic life is itself to a definite extent specialised, being a composite of nitrogen, carbon, hydrogen, oxygen, and sulphur — plastic to the vitality which is to give it embodiment and meaning. The variety of the types developed is a diversification of the organic harmony, but the most advanced organisms before at birth they emerge, each upon its own particular strain, must rise from the basic note, recapitulating ante-natally every variation of the entire gamut, each clothing itself in one singing-robe after another until it is habilitated for its proper song. This recapitulation is a successive involution rather than an evolution, no special development being allowed until the ultimate stage is reached.

Our present existence is not only an allotment in time and space, but a special allotment : every embodiment being a peculiar sequestration with fit and complementary environment. If we could see the entire synthesis in all its correspondences, the attunement would be manifest, and we would not think of one part as acting upon another, but of all as a living symphony.

VI

In an inclusion so insulated (in order that an organism may have, in any proper sense, individuality), so special, and so complex, it is inevitable that illusions must arise, enhancing the delights and deepening the anxieties and sorrows of the human pilgrimage. We have considered these illusions in a general way, but we desire here to show how directly they are associated with a definite term or cycle of existence, and especially with the apparent impotence of its conclusion. In this connection they dominate our emotional and intellectual life, and they do this by a projection which is an inversion of the living truth. This inversion begins with integration itself. That reaction which is in the expanding and ascending life, and by which it becomes an involution, we project as an external limitation. Time and space, which are only the forms of our thought, we project outside of ourselves, as if we were in them and not they in us. Resistance, which is inherent in repulsion, we attribute to outward objects. Control or restraint, essentially subjective, we regard as pressure or urgency from without. What is merely concomitant or complementary in our environment becomes in our thought the cause of states in us. We say we love what is lovable, whereas nothing is lovable save through our loving. We think of matter as eternal and of ourselves as having begun at birth—of vitality itself as something permitted for a brief period by suitable conditions of the elements. We say that life depends upon structure, and are anx-

Mental Inversions.

ious as to means of its sustentation. During a portion
of our brief cycle we grow in stature and strength and
knowledge, and in the full enjoyment of our heritage
and the widening of our perspective it seems as if life
were overflowing its bounds, and we think of ourselves
as nourished and filled from some fund provided for
this plenitude ; then waste gains upon repair, and the
wide fields grow dim and gray—when it seems to us
as if we were defrauded of all the wealth bestowed
upon us, until at last we are reduced to nakedness
and pass into the world of shadows. But, in reality,
the diminution, like the increase, is subjectively determined ; both are the visible signs of the imageless reaction of Life, which itself can be neither increased nor
diminished.

VII

In the phenomenal world, as we know it, the apparent diminution of potential energy begins with the specialisation of existence, with the divided living : it is
the sign of every beginning as of every ending, and is
more evident at every successive involution,
as in every individual organism it is con- Limitation
ab Initio.
spicuous in its special development. It is
the sign of advance in the order, in the species, in the
individual ; but because we are so sensible of it as a
barrier, in that conclusion of a term of existence which
we ordinarily call death, we associate it with weakness
and decrepitude, with the manifest impotence, forgetting that the so patent blank wall which then closes
in about us began its closure with our first moment.

Death which at last seems an intruder is in reality,
and in so far as we have part in it, what it was at our
first germination and what it has been all along—the
master inspiration, our nourishment, the storage of our
increase, our habilitation and restoration. The masque
he wears as last he looks upon us belies his mighty
office which he invisibly performs, clothing anew that
which he divests, bringing to resurgence that which
he seems to seal in with the outward hardness of stone.
Outwardly we note the final encroachment; inwardly
it is our withdrawal, the vanishing curve of our brief
cycle, a yielding to earthly elements as soft as our first
seizure upon them—a yielding which is our release,
such as we have so often had in sleep.

VIII

The sense of independence on the part of an individual organism is as illusive as that of dependence.
The harmony of the world, including humanity, con-
sists through a relation which is complementary and
not causal. The strains blend without confu-
Our Cosmic sion. The interaction between the animal
Partners.
and vegetable kingdoms illustrates this blending, as if there were oneness of action rather than interaction. It is impossible, therefore, to overestimate
the importance of environment. Human action is thus
conjoined with cosmic operation indissolubly and in
an everlasting partnership. This is as true of psychical
as of physiological manifestation, what we know as
thought having its physical side. The cosmic comple-

ment has its own reactions, apparent to us, indeed, only
in such investiture as we give them, yet inseparably in-
terwoven with that investiture—as external vibrations
are with our hearing and seeing.   In the ascent of life,
desire seems to compel its cosmic partner, as hunger
its victim, suspending that operation of physical and
chemical forces proper to them outside of this dominion
of vitality; in its descent these forces more and more
tend to resume their proper action, until finally they
bring into their own domain the structure they have
served; their hardening of the walls of life's outward
temple, begun for protection, has gone on to the extreme of fragility and destruction—an office as kindly
as any they have performed.   It is a partnership to the
very end, for while essential life can suffer no diminution, yet the individual living organism declines, the
declension being a part of its self-imposed limitation,
and to this falling the cosmic forces lean as readily as
to the rising—soon themselves to be freed from their
loving service, as is Ariel when his master escapes the
island seclusion.   The partnership is continued through
successive generations of humanity.   The descent of
the individual has, in its service of the new generation,
the aspect of a sacrifice in whose consummation Nature
officiates as high priest, burning upon altars firmly built
the last dry sheaves of the harvest.   For the passing
generation her work seems here to reach its conclusion; but she also will have her transmutations,
and meet on new terms these vanished souls.   The
descent began in the service of new life and was continued in that service; its completion is for its own
invisible ascension, as the stream, serving while it

falls, disappears only to be caught up by the sun to its hidden fountains in the sky.

IX

The integration of the individual life is a tension, an involution, a reaction and limitation. In its very expansion it becomes an inclusion and confinement, insisting upon the partial, the divided living, at whatever loss and exclusion. The simplest and most plastic organisms are inconceivably more potent than the most advanced and complex, the latter also having the greatest potency before germination, before they are aware of living. Blind feeling is sensitive to vibrations from which specialised sensation is excluded, and chemical processes depend upon solar rays more powerful than those to which any developed organism is sensible. Kinetic energy is patent through the latency of the potential. Thus the inclusion becomes an ever vaster exclusion, as if life advanced through the recession of its powers, getting its values through distance, holding its revels aloof from its central fires, distilling its dews upon the cool hard surface from which the sun has fled. The story of life is from the beginning one of abnegation. Man in his psychical progress largely surrenders the instinct common to all other animals, thus limiting his knowledge, confining it to slow and definite processes of accumulation—limiting his action also within the scope of design and invention. He seems a mere nothing in the immensity of space, and the whole cycle of his earthly

Advance through Limitation.

history but a moment in the world's time ; his work upon the earth is like writing upon the sands soon to be obliterated, and his conscious correspondences with the universe are but flashes of light in the vast darkness. Of the complex synthesis in time to which, in his present state, he belongs he knows very little, and of any other absolutely nothing. Least of all does he know himself—what was his being before he appeared in his present form, what it shall be when he is divested of that form, or even what it is now in the depths whose movements are not registered in his consciousness, certainly not registered so that he may take note of the index. Indeed, the full knowledge of any living reality would operate like the coming of Zeus to Semele, shattering his intelligence. Life so turns upon itself, in its tropical reaction, that the very terms of his knowledge change into their opposites. While he steadfastly gazes upon red it becomes green. He can make no assertion which he must not come to deny, and no denial that in its own completion shall not be confession. The trope makes the terms, and makes them those of a paradox.

But the loss is for gain; the more partial is the more complex, the divided living the field of multiplicity and variety, what is mercifully excluded therefrom permitting the express and manifold excellence of the virtue and beauty and truth of our human life; and as the contracting rocky crust of the earth is covered with tender and luxuriant growth, so man, ever at the surface of things, has there the open and extended view,

" The harvest of the quiet eye,"

the subdued melody of earth's voices, vast and intimate
communicability with things and forces tempered and
brought near, and the exquisite sensibility and motion
of the soft flesh that covers his vertebrate frame, even as
this hard structure veils the inmost plasticity of his incarnation.   The social plexus, too, above the tenacious
fabric of its unyielding laws, has the play of its gracious amenities, warm sympathies, and gentle charities.
The psychical development relieves its own inflexible
logic with the poetic dream and all the airy forms
created by the imagination ; and religious faith rises
above its firmament of creeds, transmuting the conditions of divine justice into the intimacies of a mystical incarnation, wherein it has a new motion and sensibility—the plasticity of a new principle, the oldest
of all, hidden from the foundation of the world, the
eternal kinship.   Thus the organic kingdom, ending
in man, is the reflection of the whole cosmic cycle back
to God.   It is a fleeting season, but it is the world's
summer, whose express glory is due to the veiling of
potential energy, every new limitation and hiding of
life being a fresh and more marvellous manifestation
of its creative power.

X

It is a glory that must pass, known only as it passes.
That defect, or what we deem defect, in all manifestation from the beginning, which has led so
many minds to associate matter with diabolism — that disturbance of equilibrium by
which motion is possible, so that the wheel of life may

Defect
Radical.

turn—that slight friction which, for the same possibility, science postulates as an attribute of the ether, itself
the elasticity of all tension ; all these are but other
designations for that tropic reaction of life, determining every specialised manifestation, hidden in ascent,
expansion, and increase, and disclosed in contraction
and descent. Brahma becomes Vishnu, the Preserver,
and then Siva, the Destroyer. This trope is ever
present to the mind of the Preacher : the crookedness
that cannot be made straight, a wanting that cannot be
numbered, the one event that happeneth to all, the
great evil under the sun. " To everything there is a
season . . . a time to be born and a time to die ; a time
to plant and a time to pluck up that which is planted ;
. . . a time to get and a time to lose ; a time to keep and
a time to cast away." Looking toward the inevitable
end, the view becomes pessimistic : the limitation suggests weakness, malady, and corruption, and in human
life is associated with a deeper frailty, the taint of
souls, the lapse unutterable into the bottomless pit.
" To be weak is miserable," and this weakness, this
goal of impotence so apparent in old age, when desire fails and the grasshopper becomes a burden, so
seems to set vanity at the end of things that we
wonder, in our philosophic musings, why we should
take such pains to set straight any crookedness, to
build up and buttress structures that must so surely
fall, why, indeed, our cup is filled with sweets that must
all turn bitter. The end of life thus reflects its gloom
upon the whole course, especially in the minds of those
whose hold upon existence is all along timid and feeble,
and in those ages which lack faith and vitality ; and

we almost envy that strong desire which in more primitive times led men to believe in the possibility of taking into another life their earthly possessions—wealth,
wives, and servants—that were buried or burned with
their bodies, confident, as the bees in making honey
for their winter, that somehow, though the vase of life
were broken, they might avail of its precious storage
for death's hibernation. Better still is the faith in life's
resurgence, for new increase, thus bringing us back to
the fountain.

XI

The weakness and pains of infancy are as great as
those of age : the latter call forth more of commiseration, because for them the relief is wholly invisible, and
is not ours to give; the former appeal to our helpful
sympathy, and also have help that we know
not of, even as we only partially comprehend
their magnitude. The mother knows her own
travail, but not that of her child, who never in his conscious life will undertake a labour equal to that he must
bear before he is born. Within what a brief period
does he repeat from the simplest of organic forms every stage of a development that has taken thousands
of years within its scope ! We have here in this recapitulation, this foreshortening of the work of ages, a
hint of that potential energy which is greatest in the
least specialised forms of existence—open to the Infinite. " My substance was not hid from Thee, when I
was made in secret, and curiously wrought in the lowest parts of the earth. Thine eyes did see my sub-

The Miracle of Infancy.

stance yet being imperfect; and in Thy book all my members were written, which in continuance were fashioned when as yet there was none of them."

The human germ, having accomplished its ante-natal miracle, is brought into the light of day a helpless infant; but though seeming a mere weakling, it has still before it new mountains to remove. It must wholly vitalise and bring under control its plastic embodiment; must make its connections, physically and mentally, with its natural and human environment; and, in doing this, it must supplement the subtle architecture of its brain, here again repeating in a brief period what centuries have done for its ancestors. It cannot inherit thought and speech or any experience; in all these it must begin at the beginning, and yet catch up with whatever advance has been made by its kind. Very little of this inconceivable burden can be borne for it by parents, kindred, or teachers — subjectively, indeed, naught of it; in arbitrary symbolism the signs are held out to the child, but the latter must give these their significance. The invisible power of life which shaped its organism, already limited and veiled by that organism, is still called upon to perform miracles. Outwardly there is no sign of this travail, and when it is greatest the child is nourished with milk, and spends most of its time in sleep; indeed, the tender plasticity is the essential condition of the miracle.

XII

The season of infancy has much in common with that of age, though so different are these in our thought of them. The burden of the child is invisible, not apparent in consciousness, its gravity being hidden in the expansion which is an uplifting tension ; in age the gravity is disclosed and shown as oppressive weight. The jaded sensibility of age toys with the objects of its diminished desire, simulating the dalliance and shy coquetry of the child's first contact with the world. The new desire has pain, as the old has weariness, and we see the children, thrust upon this earthly coast as by the impulse of a tide at its flood, yet crying because they have come, and seeming to question if they will stay. How coyly do they take their places at life's feast, as if nibbling at some possibly treacherous bait with dainty and quickly surfeited appetite ! Never does sweet milk sour so quickly as the mother's in the gorge of her nursling; and the regurgitation is alike prominent in Shakespeare's portrayal of the infant and in Swedenborg's vision of heavenly innocents. The unconscious desire, with its sure wisdom, though it lacks the eagerness of an acquired taste, of an appetite that has grown by what it has fed on, has yet a hidden violence ; but because the sensibility is new and fresh, its first contacts with an untried world are attended by pain and irritation and the difficulty of crudeness, as newly awakened eyes suffer the dawn, seeming to shun what they await. The bold venture is outwardly shy

Desire begins in Aversion.

and full of a play in which repulsion seems primary rather than attraction. The seizure begins with an open hand that would seem about to put aside its object before grasping it, even as it ends in relaxation, in the rejection of its fulness.

Thus, though childhood is so postulant, asking for all things, yet the first responses to its prayers are accepted with an averted face, as of those who are leaving the world instead of those who are taking it—the curvature of departure being the same at the beginning of the cycle as at the end. The cup of life has no more of bitterness in its dregs than there is in its first relish.

Novelty excites nausea as does satiety; a wholly new sensation or situation produces a kind of dizziness and bewilderment. The taste for any food, as well as for stimulants and narcotics, must be acquired, and a different zone becomes compatible only through acclimatisation. Precisely this arrangement of harmony which we enter into at birth has never been ours before, and there is a sense of discord at first and the attunement is gradual; a chaotic disturbance precedes the cosmic agreeableness. We are at first in the strange situation of the blind man whose sight has been suddenly restored—at a loss, even as one suddenly deprived of sight. Hence the feeling of sane restfulness that comes from familiarity. We are pilgrims in the far country and must be naturalised. We observe, if we do not remember, the child's timid aversion to a new face or a strange garment, and in the beginning all outward shapes are rude disguises—even all that is stimulant and helpful being first seen as hostile, and only slowly disclosing the intimate friendliness.

XIII

Pathology begins with existence, showing the aspects of malady in nascent conditions, as might be expected, since the seed must die for its own abundance.

Normal Pathology. Our physical functioning results not only in waste but in the actual precipitation of a poison, which adds malignancy to weariness. The first stage of nutrition is toxic, the stomach producing peptones, whose poison is eliminated by the liver, itself the cause of sweetness and the seat of melancholy. Even medicine relieves disease by virtue of its bitterness, and by every moment tasting death our life is forever renewed, while we smile contempt at the angel we have wrestled with for his blessing.

Difficulty, resistance, disturbance, pain — whatever names we give to the limitation upon which we enter— belong to life, to its proper reaction from the beginning, and are the basis of a normal pathology. Nascent and renascent life is in the line of resistance, is in its expansion aware of its bond, and involves disease. Comparing its repulsion to what in physics we call the centrifugal force, we think of it as resisted by attraction and as thus brought into flexion; but, really, the repulsion is from the first an attraction, and so a flexion at every point of the cycle, or vibration. The expansion involves the tension, and therefore it is that it becomes confinement.

The reaction is constant through the whole term of existence—the basis of endless change and infinite variability; forever interrupting the tendency of habit, which is toward stability, uniformity, and facility, and introducing the hostile, alien elements, dissociable for new association. For every sign of the zodiac there is some new labor; and in this travail all outward assistance involves resistance. The latent inward potency is outwardly maintained in the deepening of capacity, whose tension is buoyant, lifting as it deepens. But in the aspiration every movement is a spurning of what it meets, contempt of what it embraces, and though life makes terms with its adversary quickly, they are terms of reconciliation whose first and last import is one of disdain. We turn with weariness from Day to Night, and at dawn smite with rosy arrows the breast that has renewed our strength. The children turn against the parents, truants from home and at enmity with teachers and nurses. The Lord of Life brings not peace but a sword," setting a man at variance against his father, and a daughter against her mother," so that a man's foes shall be of his own household. Normal like abnormal pathology has its shocks and chills, its fevers and its angers—its pool of Bethesda, whose waters are troubled by the Angel of Death, who is invisibly the Angel of Life.

The strongest passion of animal life is the beginning of physical death, and we are not wholly amiss in calling its first appearance a "love-sickness," for what is there so full of pains and rages and fevers? It is the first note of command issued by That which is to Come, calling for the sacrificial festival and procession, for the Passing of the Present, bedecking every barge upon the stream with bright-coloured garlands, with music and dancing, so that no earthly vesture can vie with the gaiety of this mortal habit.

Death, as the end of life, seems especially the time
of parting ; but a closer intimacy is broken by birth,
and every crisis of our existence is home-breaking as
well as home-making. The very specialisation of life—
cosmic, individual, and social—is, as we have seen,
through division, every division or involution being a
new manifestation of reaction, and always a marvellous
surprise. In the individual the germ becomes organ and
the organ function, and so the stream runs away from
its fountain. If it were a perpetual cycle it would still
be through waves ascending and descending ; the integration being forever renewed through disintegration.
In such organic life as we know the term is limited,
with constant alternation of increase and expenditure ;
but a point is reached where nutrition is checked, and
waste gains upon reparation—the line of demarcation
between youth and age.

XIV

The burdens and pains of plastic childhood are
quite hidden, not only from outward observation but
from consciousness itself. The ascent is not like the
climbing of the Hill of Difficulty, but rather like a
translation into the heavens, the burden and difficulty
being included, as if they were participant
Exaltation of   in the exaltation, upborne by some invisible
Childhood.
power. The expansion is at the same time
a withdrawing and an imperative absorption. Hence
the quaint mastery of childhood, its native *hauteur*, its
sublime sefishness. It is said by those who have stud-

ied the child's ways of thinking, that he regards aged
people as in the state of becoming little ones. We,
on the other hand, looking at children's faces, seem to
see beyond them the abysmal realm of the Ancient of
Days. How swiftly have their softly fashioned limbs
scaled the old battlements! Ruddier and stronger than
the dawn, fresher than the spring-time, older than the
stars, they spring forever from the loins of the Eternal,
and no visible constellations may yield their true horo-
scope.

The ancient symbolism representing the apparent
movement of the sun through the twelve signs of the
Zodiac (corresponding to the twelve labours of Heracles) is true also in its application to the cycle of a
human life. First the solar hero is lifted by the help
of Aries and Cancer in his ascending movement, reaching finally the summer solstice in the House of the
Lion ; then gently declining into the arms of the Virgin, he is held for a time in the pause of Libra ; and
finally, having received the sting of the Scorpion and
the arrows of the Archer, he passes through the trope
of Capricorn into the watery region of Aquarius and
Pisces—the signs of dissolution.

The human child, like the infant Heracles, avails of
the heavenly powers with which it is secretly allied, being for a time withheld in its true kingdom, which is
not of this world. For childhood Time itself is an infinite expansion, a verisimilitude of Eternity ; the reaction of tender puissance is quick and mighty, so that
its release is as ready as its seizure, and the aged
Reaper with the scythe is not needed to make sure
the severance, as he is for them that are inveterately

14

rooted in the earthly soil. The strain of buoyancy is also its restraint, herein also showing the reaction in a sure inhibition, a marvellous continence.

Childhood, as measured by outward observation, is very brief, but in the calendar of the individual consciousness it transcends all seasons, and is indeed immeasurable. It is sacred and inviolate, guarded from the use and waste of expenditure, keeping still the secret of its deathless power, while most including and hiding death. It is a flame which consumes not—the flame of increase. The heavenly foundations are laid of life's temple, which rises like an exhalation in unsullied purity.

XV

But this wholeness is an integration which rises above ruins, and while itself inviolable is a resistless violence and ravishment. It takes all and gives nothing. Its dominion is greatest when it is most withdrawn from earthly contacts, when its walls are soft as clouds, and when as yet its voracity shows no teeth for crushing and no sting for wounding. All signs of conflict are hidden in this supreme self-centring absorption, this primal storage. The quickness of life, also, is veiled beneath the outward aspect of inertia and somnolence.

The Outward Quickening.

Achilles is still among the maidens, like one of them, and wearing their garments; the swiftness of his feet is not yet disclosed, and for him neither spear nor shield has yet been fashioned.

The time comes when the limit of capacity is reached,

when the invisible quickness becomes an outward quickening, as when the lightning that has been hidden in the
depths of the tense cloud leaps from its lair and breaks
the heavenly silence.   It is as when the bow has been
drawn to its full tension and is released for the other
half of its vibration, speeding the arrow.

We have, in another chapter, considered those "critical moments" in all development, inorganic and organic, which Mr. N. S. Shaler, in his *Interpretation of
Nature*, has treated with luminous significance.   These
belong not only to every complete cycle, but also to
every living moment, which has its two sides—of tension and release.   When the limit of tension is reached
the reaction is manifest in the abrupt action which
seems explosive in the escape.   There is this limit to
the involution of every type of existence ; and it is also
indicated in every diverse plane of the same existence
and in every particular process.   In purely physical
phenomena it is more conspicuous, as in the sudden
precipitation of a shower or in a bolt of lightning.   In
the organic world there is greater suspension and more
modulated strain.   We do, indeed, note the quick outburst of a flower, the mark of hysterical violence in
laughter and sobbing and in a passionate word or act ;
but for the most part temper disguises the tempest,
and the critical point escapes notice.   Yet every motion, every word, every thought, marks this sudden access, whereby, indeed, they become motion, word, and
thought.   There is in every process the point of abrupt precipitation, though the movement break as quietly as the surf of a summer sea, or progress in rhythmic harmony like the more distant waves, whose rupture

is hidden in their fluxion. There is the gradual reinforcement, the movement itself becoming momentum, to the
point of excess ; in youth the expenditure, or release, is
an overflow, an invisible exhalation, while the hardened walls of age resist and are broken.   In human
affairs there are crises so sudden as to be unanticipated
in the slow increment of movements leading up to them.
The masterly practical man is quick to see the first
signs of the storm before it breaks.   Hence the emphasis of opportunity, the taking of the tide at its flood.
In every great movement there is a storm-centre, toward which all the elements are drawn; the demand is
exhaustive ; it is as if the spirit of the time were mar-
shalling his hosts for an issue known only to him,
crowding expectancy, accumulating enthusiasm to fanat-
ic excess, overcharging the capacities engaged.   Then
suddenly the meaning of the movement is known, as if
certified by the announcement of angelic choirs, whose
theme becomes thenceforth the burden of human speech
and song ; the passion is expressed in the prodigality
of its blossoming, which speedily becomes the prodigality of ruin.   What matters it if the blossoms are
swept away by the wind and rain, so the fruit is set ; if
the walls of the temple fall, so the Presence that filled
the temple is glorified ; or even if the entire structure
of a civilisation is destroyed, so the race is reborn !
There is no outward explication of such crises ; it is
upon the environment that the relentless demand has
been made ; it is the external structure that has yielded
to the transformation of creative life.

Life so insists upon integration — makes such demands for it in every involution—that we come to look upon the temple, thus wondrously fashioned and at such costly sacrifice, as its end ; but the Lord, looking thereupon, saith : " Not one stone shall stand upon another." The expression of the life which shaped the structure is possible only through disintegration. Things high and holy are for brokenness and descent, whereby their essential quality is manifested. Life ascends to that point from which it may most expressively fall.

Expression in Ruin.

Childhood is the fountain in the sky, lifted thither by its vital tension, and there permitted an unadulterated storage ; in its exaltation an image of primal holiness, an unmoral innocence, not knowing evil as distinct from good. But when the time comes for it to descend into earthly channels and contacts—this is the other side of life, the contraction of its sphere, wherein it loses its translucent and crystalline purity. Yet it is at this turning-point that the individual human life enters upon its fruition, its summer, as if in the wanton prodigality of its functioning—its action and its passion—it would express all the wonder and glory hitherto hidden. It is a trope, a change as remarkable as that which befell the planet when its self-luminous orb became opaque and its barrens blossomed into the luxuriant life which expresses the flaming wonder they had veiled. Thus life falls into its special excellence, having thus also the special defects of its excellences. A special and consciously recognised pathology is developed which even in its normal course has its fevers of excess and its chills of failure. There is specific good and specific evil after the fall, and seen as distinct in a moral sense. In a period of fruition we distinguish between fruits, and guard against the poisonous; we especially consider consequences. Thus virtues are defined by ends. In a delicately poised order, of complexly interdependent relations, conscience has its culture, emphasising special control and solicitude. Prudence and temperance are appreciated as supports, maintaining integrity in a world where all things are falling and where riotous waste is so conspicuous.

XVII

As we have seen, in our consideration of the progressive specialisation of life, the suspense and temperament are more apparent at every successive stage. The species have continuance; the wave is caught in falling, and there is the undulatory procession of generations. Man dwells upon the earth, and this dwelling has new and stronger meaning with the advance of civilisation; so the moral aspect of human society is deepened from age to age in a constantly increasing conservatism. As in mechanics gravitation is made to promote levitation, so even the ruins of civilisations contribute to the greater permanence of societies that inherit their virtues. The spiritual exaltation of the Hebrew, the art of Greece, the jurisprudence of Rome, though they could not save from fall-

Maturity.

ing the structures in which they were originally en-
shrined, have become elements of sustaining power in
the structural development of modern social life.

The individual also has the advantage of this sustained undulation at the noontide height of maturity,
the prolongation of which is an extended plateau hiding
from vision the precipitous declivity. He does not see
in fruitfulness the signs of decay or how much of dominion he has surrendered for his conscious mastery.
He is not sensible of the curvature fixed by his limitation; he has the habit of walking, forgetting that there
is falling in his erect progression—the habit of speech,
unfaltering, of facile thought and action; he is conscious of rectitude, and he glories in his strength and
in the far-reaching utilities of domestic and civic functions. Like the river in the full volume of its progress,
he possesses and enriches the plain. He rejoices in
the full splendour of summer, in the decency and dignity of ample investiture. The green slowly turns to
golden, first the blades, then the ear upon the silken-
tasselled stalk, then the full corn in the ear. Surely the
value of life is expressed in its harvests, and in the
west is gathered all the wealth of the world; there are
the golden fruits of the Hesperides. These gardens
lie, indeed, on the verge of Pluto's realm; but man in
his full strength does not suspect how far the Dark
King ventures inland. The streams, of course, belong
to this invader, all lapsing Letheward, and his hands
stretch forth in the darkness of night and the chill of
winter; but Persephone, plucking flowers, found him
ere the shades had fallen upon the fields of Enna;
Adam and Eve heard the voice proclaiming him among

the trees of Eden, just in the cool of the day; and the
bright-crested aspiring serpent who had denied death
slunk away among the dry, rustling leaves to his still
confessional.   All climbing things deny him, but the
very outburst of their denial is into the leaf and flower
and fruit that in their fall shall confess him.   Yet is he
patient, letting the fruit slowly ripen.   He permits the
long-withholding of childhood from the summer heat,
waits through the long noon of manhood, and even gives
old age a staff against too swift decline.   The prolongation of maturity is itself a support to the declining
years of a passing generation, while it gives sustenance
and protection to helpless childhood and tutelage to
adolescence.

This suspense, in every period of human life, emphasises the value and importance of that life, considered
solely in its terrestrial relations.   Mr. John Fiske, in
showing that the prolongation of human infancy has
been one of the principal factors in the progress of the
race, made a novel and original contribution to the science of sociology.   But if the weakness and dependence of childhood, evoking loving care and sympathy,
counts for so much, how much more must be accred-
ited to the invisible might of childhood as the hope of
the world.   During this period of protection, while it is
establishing its cerebral channels of communication
with the outside world, it is at the same time, by its
withholding from that world, allowed freedom for expansion, for the deepening of its capacity, for that exalted tension which society has come to recognise as
the mightiest of its inspirations.   This mystical apprehension of childhood becomes the poet's assertion and

the popular intuition; and, since it regards elements not open to observation, it is a view falling outside the scientific scrutiny that regards only the stimulation of environment, the nutritive processes involved, and the resultant structural development. "What is this wondrous font of power?" asks science. "Is it anything more than a fund of vital energy dependent upon nutrition for its storage?" In return, we ask, what is it at any stage of its outward development? At what point in the stream does this transcendent, invisible power which gives human life its spiritual meaning enter, if it is not at the fountain? It is not an acquisition. If we admit it into our view of human existence as a whole, we must include it from the beginning.

Indeed, as we have seen, this involution which we know as childhood is at the fountain something that it is not in the stream. Its expression is also its veiling. "It is not as it hath been of yore," the poet complains. A glamour is gone that never comes again, it

"... fades into the light of common day."

The virginal sense of things first seen; the surprise of fragrance; the native feeling of primal dawns, of the heavenly azure, of woods and streams, of haunting shadows and whispering winds, we cannot recall. The steps that halted then are hurried now, following well-worn paths and yet lost in them. The storage of strength against strain, of reparation against waste, is not like that primal storage, which had its basis in a hunger that was not want. No after-sleep is like the sleep of the infant, which is not measured to meet a special weariness, but is rather the sign of the hidden

quickness of life in its infolding, as wakefulness is of
the quick unfolding, growing into the insomnia of old
age.   Yet the nutrition and sleep of adolescence and
maturity are special infoldings, whereby the haste of
the consuming flame is retarded and the plasticity of
childhood is in some degree renewed, though it cannot
be wholly regained ; and waste and weariness induce
and stimulate these processes of renewal.

This period of maturity, sustained by constant reinforcement of energy, is far remote from childhood, but
it is true of the man as of the youth, that he, though he

> ". . . daily farther from the East
> Must travel, still is Nature's Priest,
> And by the vision splendid
> Is on his way attended,"

and this vision illumines his ripe knowledge and gives
its own transcendent meaning to all he does.

XVIII

The suspense is in some measure maintained in the
period of decline.   The urgency of physical passion is
spent and the intense strain of effort is relaxed ; in the
golden silence, beneath all the easy garru-
Decline.
lousness, contemplation is deepened, undisturbed by passionate interest.   The last juice expressed
from the vine is unutterably rich.   Memory seems
weaker, but it is busy at the old font.   The flame of
life which burned only green in the spring-time bursts
forth into many brilliant autumnal colors, as if death

had more gaiety than birth. Age seems to be a taking on anew of childhood, but with this difference—
that the reaction awaits some other sphering of the
withdrawn life.   Instead of the aversion which ends in
seizure there is the lingering clasp of cherished things
about to be released—love mingling with the weariness, so that the final human repentance of the visible
world is unlike that of any other species in its regret-
ful, backward glance of farewell.   In man alone does
love conquer the strong animal instinct which insists
upon solitude and utter aversion of the face in death.

XIX

The urgency of the movement, hidden in the ascent
of life, is outwardly conspicuous in the descent.   There
is more of death and destruction at the beginning than
at the end ; the unconsuming flame is most intense,
though there is no smoke nor conflagration.

It is with Death as with Evil—neither is   *The Disarray.*
apparent to us, under its name, in the uplifting tension of life, which most completely includes
both.   The flame is tropical, and when it turns it rends ;
its reaction is disclosed as a wasting consumption.   In
all germinant organisms we note the hidden quickness
of the tender infolding life, and in the unfolding an
outward quickening in blossom and song and radiant
plumage, when, with the prophecy of new life to come
in the ripening grain, the fertilisation of flowers, the
mating of the birds, and the myriad forms of love-life
in the whole realm of animate Nature, another move-

ment begins, hurrying into flight, which comes at length
to have in it a suggestion of disorder and disarray. The
song sung by the weird Sisters, when they unravel
their slowly woven web, has reckless, dissolute notes.
The ascendant movement of life, with its hidden quickness, its virginal restraint, seems outwardly slow, and
has outwardly also the aspect of ease and buoyant rest,
because the travail of its climbing is mainly borne by
unseen powers; but, in the descent, it would almost
seem that these benignant powers, breaking through
the veil, had suffered a transformation and become destructive foes, losing their coy reticence and playful
ease, and were striding forth in open, undisguised violence, and with indecorous haste were flinging their
garments to the winds, bringing all things to nakedness, profaning all shrines, ravishing all Beauty, brand-
ing Plenty as wantonness, and Accomplishment as vanity. What was, in the nascent organism, abundant,
graceful ease and rhythmic overflow, nourished from
hidden sources, becomes, in the decay of the organism,
a feverish excess, a hectic waste. It is the trope of
Capricorn, and the pagan imagination was easily in-
fected by its disturbance; the followers of Pan clothed
themselves with goat-skins, and grinning satyrs mingled in the wild rout.

For man as for all other organisms there is, in the
visible course of things, the lax and ragged conclusion
—the broken golden bowl at the fountain and the
wheel broken at the cistern. The fountain cannot refuse to become the stream, nor the stream to pass;
any arrest of the descending movement only accumu-
lates disturbance and hastens the ruin. It is the bitter-

ness of Dead Seas that they have no outlet. The unbroken storage of the miser becomes itself corruption.
The belief of Heraclitus in the eternal flux of things
must somehow be reconciled with Plato's plea for stability through a harmony that is eternal.

There is no ethical resolution of the problem; there
is indeed no problem save of our own making. The
issues of life have their spontaneous reconcilement, because Life itself is eternal. There is in that life a
principle which is creative; which is as unmoral as
is Childhood, because it transcends morality; which
makes not for mere rectitude, but for righteousness, not
for betterment merely, but for renewal; which does not
mend the Prodigal's rags, but brings him home.

FOURTH BOOK

DEATH UNMASQUED

CHAPTER I

A SINGULAR REVELATION

I

IN every system known to us some singular and striking phenomenon presents itself — a certain insistent strain of the harmony, not easily explained, and in many cases remaining forever an insoluble mystery. The Milky Way, the ^{The Broken Man.} Gulf Stream, the Trade Winds, the current that rules the magnetic needle, are such phenomena in the physical world. In physiology the quickening and dominant power of germ-cells discloses to the student the plasmic Milky Way of organic life. The Dream impresses us as a similar mystery in psychical manifestation. Thus singular and inexplicable, in the currents of human history, is that one of them which determined the Hebrew destiny.

The Gentile, or pagan, races of the ancient world accomplished outward integrity, or completeness, in the development of art, science, and polity; they had humane literature and elaborate religious ritual. The Hebrew was pre-eminently the broken man. Those prophecies which we usually regard as wholly Messianic were first of all applicable to Israel. He is the one spoken of by Isaiah as " a root out of a dry ground :

he hath no form nor comeliness ; and when we see him there is no beauty that we should desire him. He was despised and rejected of men ; a man of sorrows, and acquainted with grief. . . . He was wounded for our transgressions, he was bruised for our iniquities : the chastisement of our peace was upon him ; and with his stripes we are healed." All this, consummated in the person of the Christ, pertained to the race whence he sprang. For the Hebrew the promise of the rose presumed a desert. "Look unto the rock whence ye were hewn," says the prophet, "and to the hole of the pit whence ye were digged. Look unto Abraham, your father, and unto Sarah that bare you." In Abraham's seed all nations were to be blessed ; but how suggestive in this primitive gospel is the emphasis upon the sterility of Sarah, and, after the birth of Isaac, upon Abraham's renunciation of him, completed in the heart, though the hand stretched forth to slay was stayed !

Always in the history of this race, despised above all others yet above all others glorified, Canaan must have its prelude in the wilderness ; some bitter tribulation like that of the Egyptian bondage lies ever in the background. Canaan itself—the land flowing with milk and honey—was a field of terrible carnage, possessed only after many fierce battles, and with difficulty maintained, lying between Assyria and Egypt as between an upper and nether millstone. Its captive children were sold in every slave-market of the Mediterranean. The kingdom established by David was short-lived ; in the generation succeeding Solomon it was broken in pieces, and ten of the twelve tribes soon disappeared so com-

pletely from view that their fate has become a historical enigma. The remaining tribes of Judah and Ben-
jamin, harassed for several generations by foreign and
intestine wars, were carried away in captivity to Baby-
lon, from which a small remnant returned to rebuild
the ruined temple and rescue from oblivion the precious records of the past.

It was after this captivity that the more gracious aspects of the Mosaic law were emphasised, and there
arose the sect of the Pharisees, in its origin representing
the loftiest spiritual ideal ; for the first time formulat-
ing the belief in a resurrection ; and, in the institution
of synagogue worship throughout Palestine, establishing the simplest form of religious liberty ever known
upon the earth.

After every black night in Jewish history there was
some such glorious morning. It is true that in our
Lord's time Pharisaism, especially in Jerusalem, had
degenerated into a habit of formal righteousness, but
the simple religious life in the country villages was to
some extent maintained, and here it was that the mere
remnant of a remnant awaited the blossoming of a
people's hope.

At the birth of Christ his country was reduced to
the position of an insignificant province of the Roman
Empire, and his people were dispersed throughout the
then known world. Into the deepest darkness shone
the star of Bethlehem.

Other races seem to have grown corrupt within their
outwardly completed structures. The Hebrew, outwardly broken, was inwardly made whole in the beauty
of holiness. Many were called but few were chosen ;

and it is not strange that in this trial by fire there was so large a refusal of dross, and that only in the hearts of a faithful few was a destiny so singular maintained and cherished to its final consummation.

II

Looking back from the eminence of our Aryan civilisation, and considering what different races have contributed thereto, we behold this one vaulting, flame-fretted arch, distinct from and overreaching all others.   It is a sacred flame—how dread even to the Hebrews, who in the wilderness saw it as a pillar of cloud by day and a pillar of fire by night, and would fain have fled from its awful illumination back to the flesh-pots of Egypt! With what natural yearning toward some familiar human imagination they moulded the golden calf, even at the bidding of Aaron, while Moses was with God in the mount in the midst of the cloud which was the glory of the Lord, and the sight of which was like devouring fire.

The Sacred Flame.

Repellent also to all men is this sacred flame, and it is with serene satisfaction that our Western thought turns to "the glory which was Greece and the grandeur that was Rome"—to those elements in the fabric of our modern life which are of classic origin, and which commend themselves to our esteem as associated with æsthetic development, with intellectual culture, with ethical stability, and with the pride of human accomplishment, attested by monuments whose ruins seem to us more hospitable than do the tents of Shem, or that holy tabernacle built by the descendants of the Bedouin patriarch, in which dwelt the flaming Presence.

Nevertheless, this arch of fire transcends all others in our spiritual temple, surpassing all earthly splendours; it is the illumination of our heavenly heritage, from a promise uttered to man in some earlier and deeper sleep than fell upon Abraham—a promise answering to the inmost desire of the human heart. The outward aversion from it has recourse in an irresistible attraction thereto. The glory of the Lord, shining in another face than that of Moses, subdued all hearts, and the world eagerly ran after that from which it had seemed to be running away.

III

The tendency toward structural completeness is natural and wholesome ; it is development in human existence as it is in the entire cosmos. It is itself a breaking, but a breaking into wholes, *Principle of Divestiture.* even in the minutest molecules. The fractions of living Nature are themselves integers. Form and comeliness are cosmic distinctions. The bride is arrayed for her lord. The lack of proper vestment, like deformity, is a cause for shame and disappointment. Nakedness is clothed upon. The more sacred the flame, the more carefully it is hidden, and the holiest passion is veiled. Life's revels are masqued, and the vesture is manifold ; this is the way of all prodigal sons, yet the fact that it ends in raggedness and ruin is Nature's confession that the Life is more than meat and the body than raiment.

This truth which Nature confesses at the end of things, *in articulo mortis,* the Lord disclosed at the fountain, as the spiritual principle of life. Thus was the inclusion of death in life illustrated, in his personal career upon earth, by his denial of those things which in the natural course of human lives are accounted most desirable. He renounced without denunciation. He never married, but marriage he blessed. He sought not earthly honours, possessions, "troops of friends," but to these in themselves he attached no blame; he counselled his disciples to make friends of even the mammon of unrighteousness. In saying that Mary had chosen the good part there was no reflection upon Martha. He was not an ascetic; his very divestiture was abundantly vital. As Nature, insisting upon death, yet values not the waste and ruin but rather refuses them, driving them out of sight with the violence of her winter winds and utterly consuming them in the white heat of her frost, so the Lord reckoned not with the dead while he glorified Death. "Let the dead bury its dead." He was at one with Nature, who lays such emphasis on death, because through death is her resurrection; but the truth in his word was a spiritual principle transcending that expressed in the apparently closed circles of all natural procession; it revealed the reality hidden beneath the appearance from the foundation of the world.

IV

In the natural course of things man sees good and evil apart, taking the one with delight, succumbing to the other as inevitable. He rejoices in the morning, but night wins acceptance because of his weariness, which is a kind of forced repentance of the day; and the deeper night of death over- *Natural and Spiritual Repentance.* powers him in the same way, so that he seems in a natural repentance to turn from the world to his confessional. He is overcome of evil. But the Lord said, " Be not overcome of evil, but overcome evil with good." Again he said, " Resist not evil." Now, he well knew that as in time past so in all time to come the phenomenal conflict with evil must continue. In the prayer he taught to his disciples were the petitions " Lead us not into temptation, but deliver us from evil." He was not enjoining upon men, in the practical conflict of life, to confound evil with good. He might as well have bid them confound light with darkness. Tares were not wheat, though they grow together and must continue to grow together until the harvest. What he announced was a spiritual principle touching the reality beneath the phenomenal struggle. It is as if he had said : " Evil and Good as seen by you appear separate and irreconcilable, because of the limitation of your vision and of your existence ; your thought and care and effort are engaged in a conflict whose terms and conditions you cannot evade, and yet no man by thinking or striving can add one cubit to his stature ; the visible limitation remains ; and the conclusion of the struggle is the apparent triumph of Evil—even as the grave swallows up all that live and Death seems the Conqueror. In this partial view, this finitude, this closed circle which you call the course of nature, you are like prisoners and captives, accepting evil as slaves accept the lash of a taskmaster. But I show you a hidden truth, masqued and disguised by visible Nature—a divine way, whereby as children and not as servants you shall accept Death and Evil, including and comprehending them in that true knowledge of the Father and the Son which is eternal life, in its spiritual meaning. Outwardly there is the striving in narrow ways, seeking ever narrower and straighter, but inwardly there is peace and reconcilement. This is faith in the abounding life that forever springs freshly from its fountain ; herein is the willing repentance that is not mere weariness — the losing of the soul to save itself, the taking of the yoke to find it easy, the drinking of the cup to its dregs to taste in these its sweetness. The Pharisee comes to the temple and offers up to God his righteousness ; the publican comes and offers up his sins—in him is repentance possible, a complete burial, a new birth. A man may strive outwardly against evil in every shape it outwardly takes, and yet so know the Father that he shall see that against which he strives as something essential, lying at the very root of life—that his open adversary, stripped of his disguises, is invisibly his friend from the beginning. And, again, a man may strive and trust alone to his strength, seeing good and evil only in their disguises, and for a season he may accumulate the good and fortify himself against the evil, securing comfort, safety, and outward integrity; yet shall the inevitable end come when the edifice shall be broken up and its treasure be found corruptible, having no heavenly root or lodgment. Evil is known only as an enemy, and Death as the last enemy; the adversary is never seen as the friend—there is no reconcilement. The whole need not a physician, but they that are sick. Blessed therefore are the meek in their expansive heritage; blessed they that take to their hearts grief and poverty, hunger and thirst, and desolating defeat—for in all these they shall know Evil and Death for what they truly are in a divine Creation."

V

But the Lord did better than say all this: his life was this eternal truth incarnate. He "became Sin" and glorified Death.

The imagination which created the legend of the Wandering Jew, upon whom, in the presence of a divine death, fell the doom of deathlessness, introduced into the scene with which it was associated an element of striking contrast, suggesting the beatitude of mortality at the moment of its brightest illumination. Even as contrasted with Evil and Death not thus divinely illustrated, no more dreadful sentence could be pronounced upon any child of Earth than this: that for him there should never be pain or sickness, any hunger or thirst, any shadow to break the endless continuity of light, or any death. To make utterly impossible any benediction, to this existence upon ground not accursed for its sake, it would

Christ Glorified Death.

only be necessary to add to the sentence its awful concomitant: Thou shalt never fall. Atropos, the un-
turning one, could take no surer shape. The fixed
horror of such a fate, so suggestive to us of utter
weariness, would in reality lack even that relenting in
its motionless apathy. But in the presence of the meek
and lowly Jesus, bending beneath the weight of the
cross, the blank, inflexible doom becomes unutterable
and unthinkable, until the imagination of it vanishes
into absurdity.

For, behold, the Lord had fallen ! He had descended from the bosom of the heavenly Father, and all his
life upon the earth had been downward—away from
the rich and powerful and wise and consciously correct
to the poor and sick and sinful; and now this descent
was to be completed, in the grave, even in hell—from
the zenith to the nadir. Lucifer no farther fell, nor any
son of Adam following him, than did this second Adam—
Life-bearer, but drinking all of the mortal cup ; Lighter
of the Way, but taking all its darkness, even the mid-
night of its lowest abyss.

Thus was Death illustrated and made glorious, showing at its core, its sting having been taken, a strange
and mystical beauty, not hitherto suspected, and not
apparent in the shining perfections and accomplishments which men reach after all their lives. The
Lord's blessings had always been upon the victims in
the strife of earth, and in the most human of his parables he had shown how the returning prodigal had been
given the best robe and the merry-making feast—signs
of a loving father's rejoicing that aroused the envy of
the unroving elder brother ; and in many ways he had

taught the preciousness of lost things found, the glory of defeat turned into victory. Now he was about to make death itself enviable, so that men would run after it, fearful, indeed, lest they should escape martyrdom — so that they would listen with delight to the prayer for the passing soul, invoking the sure and speedy work upon it of purgatorial flames, expecting that way some secret excellence.

VI

But the Lord did not teach men to seek that which we commonly call death any more than he taught them to do evil. It is true, moreover, that he saw no living righteousness in what men call good — in conduct having reference to those particular ends which men seek as children of this world. He revealed to men a larger heritage, an eternal kinship—they were the children of God. He referred them to this fountain of love and light, from which every human heart had its pulsation, as the well lives from its spring. To be born again was to know one's self as a child of the Father — to know and do His will. As children of this world, men distinguish between good and evil, and so, under their limitation, they must, knowing benefit and harm from their relations to a system which has beginning and end ; but a new birth brings a new vision, wherein it is seen that God creates Evil as He creates Good, and that, as parts of this Creation, they are complementary.

Inclusion of Evil.

Is not Christ the Word from the beginning, and so Nature before he was the Christ—including all that in

Nature we call evil as well as what we call good? He was the first Adam as he is the last; as the first especially the son of God, and as the last especially the son of man. Thus twice humanly incarnate—the root and flower of the race—he is truly the Head of humanity, identified therewith from the beginning even unto the end.

He never blamed men for their failures or praised them for their goodness, because he knew the limitation of every creature. From the heart of man, as from the source of all life, proceeded both good and evil, but in the new heart—that of the child born of the spirit, and seeking perfection not after an outward pattern but after the divine type; that is, to be "perfect as your Father in heaven is perfect"—the good included the evil. This is the reconcilement. To do the will of the Father, life must be willingly accepted on its own flaming terms, including that which will ultimately burn away all its outward vesture—even its habit of goodness.

Is not this to bring man into harmony with Nature, in all whose cycles of motion, truly seen, repulsion ends in attraction, and is really one therewith from the point of departure?

The limitation itself is a bond of return. The place of exile is sure to be home, and existence in time has its ground in the life eternal.

VII

Why do we think of Christ as the Eternal Child? And why did he present childhood as the type of the kingdom of heaven? The child is unmoral, has not discretion or prudence, and is not guided by the maxims of experience. These are negations perti- nent to a spiritual life in its latent powers; but positively childhood represents the prin- ciple of such a life, because in it evil is hidden, as, indeed, goodness is also; its germinant and expansive life is an ascent upon the wings of death. It is the season of taking rather than of giving, when capacity is deepened; when, at the same time that it is most energetically making its connections with the outside world, it is most withdrawn from that world, its communications with which are wholly for its own sake, availing of the descending min- istrations of other life for its own ascension. When, at a later period, it knows self-sacrifice, then is the abun- dant death it has taken given up, yielded in expenditure, becoming patent.

It is this uplifting power, making death and evil its ministrants—wondrous in its growth, in its vitalisation of its plastic organism, and in its supreme elasticity; quick in its reaction, so that no possession clogs or encumbers; the fittest symbol of creative might and authority—which the Lord had in view when he made childhood the type of his kingdom. In that view also was comprehended the unhesitating trust of the child and his fearless meekness and docility. All these qualities, in their heavenliest excellence, are combined in our conception of the Christ Child.

VIII

But childhood is the type only; that which it represents is a fuller expression, with deeper meanings. The childhood is continued into manhood in the Christ-life—into the expenditure, the sacrifice, the descent, and yet in these maintaining the type. The latent potency is developed, but still keeps its plasticity, through a willing surrender of all those outward things which, in the ordinary line of human experience, make manhood desirable. The exercise of power in this line was suggested to the Lord in the temptation on the mount. To the child the possession of earthly things has little meaning; he accepts all gifts as toys and falls asleep among them, showing instinctive contempt of those functions and uses familiar to mature experience; but to the man the offer of external grandeur is the great temptation, and he may yield to it legitimately with the determination of a righteous exercise of power, truly magnifying his office. Therefore when Christ puts aside the temptation it means more than the instinctive contempt of the child; it is a willing rejection. It is something, too, quite different from what is commonly called self-denial; in the course of ordinary experience, the acceptance might be the true altruism. The Lord would have rejected the office of High Priest of Jerusalem as readily as he did that of King of the Jews, which the people expected the Messiah to take. It was officialism itself, whether sacred or secular, that he renounced. He refrained from entering into those domestic relations

This Type as Developed in Christ.

properly enjoined as duties upon a citizen of this world. Because he was to be the real priest and king of all men, because he was to illustrate man's divine sonship, he repudiated for himself the insignia of a power and kinship which meant less than these. The renunciation was a sacrifice only in the meaning expressed by the Psalmist: "Lo, I come to do thy will, O God." In the worldly view this withdrawal from benefits ardently sought by all men, and from duties held to be most binding and sacred, seems to be an anticipation of the divestiture wrought by death. In reality it is the introduction of a new death, bringing it next the new birth. It is a natural intimacy, repeating the process which goes on in the germination of any seed, the outward husk of which is dissolved for the abounding of the inward life: in another sense it is mystical, since the new life is drawn from an invisible fountain. It is the abundance rather than the divestiture that is the spiritual reality. As in childhood, so in all germinant life: there is a hidden violence, an immeasurable might, something imperative, which makes a kingdom. In the Christ this is marvellously shown in the multiplication of the loaves and fishes under his dividing hand, and in the healing virtue of his touch. His growth to manhood is described as a growth in grace, keeping the plastic and creative potency; and that all evil as well as death is solvent at this fountain is aptly expressed in St. Paul's saying that "Where sin abounded, grace did much more abound."

IX

We see, then, why loss is the first word of the kingdom of heaven, and why the baptism of the Lord is with fire. It is because flame destroys that it is constructive; and this thought brings us back to the Hebrew, and enables us to better comprehend his outward brokenness and divestiture; for the flame which in the Christ was the illumination of the spiritual truth of an eternal life; which in its fusion absorbed and consumed the external fabric of exist-
Child Type Developed in the Hebrew. ence — the habit which men called good as well as that which they called evil—and which became the pentecostal flame of a new human fellowship, was the consummation of that which burned in the heart of every faithful Hebrew from Abraham to Simeon—a torment without, but an inward peace.

In many ways the Hebrew race, in the fulfilment of its peculiar destiny, foreshadowed the spiritual principle illustrated in the singular life of Jesus. As he was the Desire of all nations, and therefore could not mar his brightness as the Sun of a spiritual system embracing all humanity through any merely worldly aspiration, so the promise made to Abraham was one including all nations, and this large expectation would have failed of its true expression in earthly successes and triumphs, in the attainment of those things "which the Gentiles seek."

We think, too, of the ancient Hebrew as a child, and in a peculiar sense the child of God. "The Hebrew children " is a characteristic phrase, as applicable to a people always in a comparatively plastic state, and whose language never departed from its native and radical simplicity.

Considering what the spiritual life of the Hebrew means for us, we are surprised that a vine which has spread over the earth occupied so small a garden in its original growth, quite escaping the notice of classic history. In no field of human achievement has the ancient Hebrew left any signal monument of worldly grandeur. We can account for his political insignificance by situation and circumstance, but for his lack of any positive accomplishment in science, art, or philosophy we can find no explanation save in his peculiar genius and destiny; and of these the only ancient sign left us is his sacred literature. That he was not destitute of imagination is shown in this literature, which is as singular in its distinction from all others as was his whole history from that of all other peoples. Here the imagination takes its loftiest flight in song and prophecy, and its simplest strain in the quaint records of patriarchal life, in the story of Joseph and of Ruth, and in the most fully incarnate idyl of passionate love ever put in words —the Song of Solomon ; and there is no appearance of incongruity in bringing together all these into that sacred collection known to us as the Holy Bible. Never in human expression has there been so intimate association of the sensibility of the flesh with the highest spiritual exaltation ; and we note the absence of that which lies between the spirit and the sensibility—that play of mental activity which is so especially the charm of almost all classic and of all modern literature. In this

16

connection, it is significant that while the Hebrew gave
a natural expression to his emotions in the song and
the dance, and delighted in personal adornments, in per-
fumes and savory foods and wines, bringing these also
into close association with religious worship, he had no
representative arts, such as painting, sculpture, or the
drama.   While his spiritual expression was thus so directly incarnate, he did not seek that perfection of bodily
exercise which, among the Greeks, was the result of
elaborate athletic training.

It may be said that this lack of completeness is explained, as in the case of any barbaric race, by the fact
that the Hebrew was so backward and unprogressive—
so slow to put away his childhood.   But this is his very
singularity.   Why was he thus withheld in the plastic
state of childhood?   It is not true of the Hebrew race
that it was barbaric, in the proper sense of the term.
Other Semitic peoples from the same old Arabian desert,
like the Phœnician and the Assyrian, were builders of
cities, and advanced rapidly from the nomadic state into
very complex forms of civilisation.   Others still remained in the desert, where they may be found to-day,
degenerate, indeed, but otherwise living in the same
manner as did their ancestral tribes four thousand years
ago.   During the long period of the patriarchate, which
was a prolonged childhood, the spiritual capacity of the
Hebrew was deepened; but the quality and might of
this expansion are indicated and measured by the resultant movement, culminating in the appearance of
the Messiah and the resurrection and revitalisation of a
dead world ; and they are not to be accounted for by
any outward condition.

The prolonged childhood was an essential prelude to a so singular manifestation : it was a childhood maintained after the disappearance of the patriarchate, and through the entire cycle of the Hebrew destiny. One of its characteristic traits is shown in the wonderful power of assimilation. It has been asserted by profound scholars that the Hebrew derived his Sabbath from the Babylonian, the institution of the Judges from the Phœnicians, and the rite of circumcision from the Egyptians, along with the ark, the Shekinah, and the *Neshulon*, or brazen serpent, which held its place in the Holy of Holies until it was thrust out by Ezekiel. Even his idea of angels and of a future life is said to have taken definite shape through contact with the Persians after the great captivity. Assuming that all this is true, it would only show the marvellous selective genius of the Hebrew. Does the child prepare for himself his heritage? He accepts that which he has not made, but he makes it his own, and from his own heart gives it a meaning. The purpose involved in the spiritual destiny of the Hebrew "is purposed upon the whole earth ;" therefore to this child the earth is a heritage, and the whole world brings its offerings. What, then, if the skilled men of Tyre built Solomon's temple? In Isaiah's forecast of glorified Zion the stranger's willing tribute to that glory is magnified. "The Gentiles shall come to thy light, and kings to the brightness of thy rising. . . . The abundance of the sea shall be converted unto thee, the forces of the Gentiles shall come unto thee. . . . And the sons of strangers shall build up thy walls." This Hebrew childhood stands for that of humanity—its issue is the Son of Man.

This people was by its fervid enthusiasm lifted to a plane of expression so lofty that its pride was not in the initiation of institutions any more than in their perfection ; only that inward grace was regarded which gave them a living soul. Its possession of outward things was an adoption in the name of the Holy One. The zeal was also a jealousy. Whatever hands raised the temple, the Jews would have destroyed the edifice rather than admit within its sacred enclosure the statue of a Roman emperor. The attempt of Antiochus Epiphanes to merge Hebraism into Hellenism aroused the heroic and successful revolt of the Maccabees. In the early period, when the patriarchs in alien territory recognized the power therein of alien gods, the jealousy of a tribal religion was consistent with the tolerance of other religions equally provincial, and was very different from that which in later times guarded a comprehensive faith in a Jehovah who is the God of all the earth—this guardianship implying a responsibility as broad as the faith. In this higher view, Israel was a peculiar people, not as one enjoying exclusive benefits, but rather as undergoing special sufferings for the whole human race—a view not easily maintained save by the very elect, but cherished by the prophets in every age.

The divestiture of the Hebrew was as conspicuous in his religious as in his secular life. He was forbidden to make an image or likeness of anything in the heavens or in the earth or in the waters under the earth. Every other prohibition of the Decalogue was deemed as obligatory in the Egyptian system of ethics as in the Mosaic law, but this was distinctively Hebraic. In

their beginnings the arts of painting and sculpture
have always been associated with the expression of religious feeling, but they were denied any nurture by the
Hebrew faith. The prohibition is not merely the exclusion of polytheism and idolatry, but of all representative art. A living movement must in no way be arrested in a dead thing. The swiftness of the primitive
paschal feast, the erect attitude of the participants suggesting expedition, showed the indispositon to loiter in
any sacred way. The prophets always regarded with
aversion the elaborate ritual of the temple worship at
Jerusalem—a living movement arrested in fixed forms.

Symbolism was not excluded by the prohibition of
the *simulacrum;* rather it was heightened, keeping more
closely to an inward meaning. The one essential divine symbol was man himself, God's express image in
the world of living things. The Hebrew progression
in spiritual lines was toward the God-man, it was the
culture of an Emmanuel.

The human nature of the Hebrew was the same as
that of every other race, having the same aspirations,
mental, moral, and religious, the same eager desires
for earthly possession and power — for all, indeed,
which it seems to have been denied, and these natural
tendencies common to all mankind were not only amply
illustrated at every period of this people's history, but
intensified by unsatisfied hunger. The great majority
fell away centuries before the appearance of the Messiah, drawn almost irresistibly by the fascinations of
the pagan world—its nature worship, its indulgence of
fond imaginations, its splendours and dramatic pomp;
and of those who were held to the lofty strain, how

many were hedged in by the compelling Angel of the Lord or subdued by suffering and the pressure of circumstance; how many were alarmed by the threatenings or persuaded by the pleadings of the prophets! But to the faithful few who waited for the glory to be revealed—to the seers and the prophets and the guileless country shepherds—there was another charm, more potent than any which could appeal to the sense or the intellect—the charm of that expectation which lifts the heart of the mother waiting her time, radiant in her travail. Here was that Israel which should "see of the travail of his soul and be satisfied." Here burned that sacred flame which preyed upon and devoured the embodiment.

X

To the early Aryan also God was a fire—a fire which built and beautified the world; which was the fervour of the animal and the glory of the flower, and which had

Distinction between Hebrew and Pagan Faith.

its intimate human symbol in the flame upon the hearth-stone, the centre of familiar affection and loving kinship. But to the Hebrew, his God was a consuming fire, which so *rebuilded* life in a new heaven and a new earth. Where the pagan saw creation with its ceaseless round of birth and death, the Hebrew with prophetic vision saw recreation—a new death and a new birth. "Art thou a master in Israel," said the Lord to Nicodemus, "and knowest not these things?"—the things pertaining to the mystery of regeneration. The charm of such a faith is that of a desire never exhausted in outward realisation, and so conserving its native might. In paganism the religious instinct was given complete scope. Paul complained of the Greeks that they were too religious, and he welcomed the signs of a worship of the unknown God, of a divinity not circumscribed by the limits of imaginative definition and of ritualistic familiarity. The pagan system of worship was a network of ritualism and a hotbed of sacerdotalism. In its beginnings, true to Nature, the lines of its development were brought to completion within the closed circle of a visible environment, so that the secret of Nature itself was hidden. The Hebrew faith looked forward to the divine-human Incarnation; the pagan anticipated this incarnation, exhausting its imagination of it in types which fell short of and precluded the transcendent intuition.

XI

The conservation of the spiritual principle through the incompleteness of outward form and structure was promoted by the Hebrew prophets. Whenever the race was borne aloft in the common aspiration of all civilised peoples for military glory, for the luxury and grandeur of cities, for the splendours of a royal court and a temple ritual, it was continually thrust back to earth, prostrate as one possessed by demons, and by prophetic exorcism was compelled to confess its peculiar destiny. These prophets were thorns in the flesh of kings and of priests; they were the great disturbers, the preachers of humiliation; but they were the people's hope, and though in their

Mission of the Prophets.

sadly triumphal journeys they rode upon asses, they
were hailed by popular acclamations and recognized
as pre-eminently men of God. Through their influence social ambitions as well as national aspirations
were held in check. The Prophet was ubiquitous and
irrepressible, and from the time of Samuel there was
a school, a continuous succession, of these witnesses
to a Lord surely to come on earth. They remoulded
sacred traditions, and the critical scholar detects traces
of their illuminating and transforming influence in the
pages of holy writ, giving a deeper meaning to the
record of creation, the legend of Eden, and the summa-
tion of the Law.

XII

To bring in the Eternal Child, and to show that in
him man is, even within the limitations of time, the
heir of an eternal life, was the Messianic destiny of the
Hebrew. This plant in the garden of the
Hebrew Thought of Lord was diligently tended by the divine
God. husbandman, relentlessly pruned, cut back to
the quick, and thus was ever kept green and tender, as
on the very brink of an exhaustless fountain.

Often the vine strayed beyond the garden wall and
lost its succulence; perversions there were to which
even the prophets were reluctantly indulgent, as in the
popular clamour for a king; the law, so gentle in its
spirit and associated with the meekest of men in its
beginnings, came to tolerate a kind of rigid justice —
the requirement of an eye for an eye, a life for a life;
and, according to Ezekiel, the perversion was some-

times of divine origin, Jehovah himself giving his people "statutes that were not good and judgments wherein they should not live," and "polluted them in their
own gifts, in that they caused to pass through the fire
all their first-born, that He might make them desolate . . .
to the end that they might know that He was the Lord."
This declaration, so startling to a modern ear, was not
intended to convey the impression that God tempted
men to do evil, but was a forcible expression of a conviction, characteristic of Hebrew faith, that the responsibility for evil as for good was in the largest sense
divine. "'I create good and I create evil,' saith the
Lord." It was not permitted the Hebrew to think of
his sins as his own. His derelictions were monstrous,
and he needed the prophetic consolation that the Father shared the wanderings of His children, encompass-
ing them in infinite wisdom and compassion, so that in
the end they might see that the way of even the widest
wanderer was the way home.

Thus the flexibility and plasticity of the Hebrew
childhood was maintained even in his idea of the law.
Through the Pentateuch runs the warm current of
divine tenderness, in its merciful intention including
also with the children the stranger within their gates.
It is a protest against inhumanity of every sort. In
no sacred scripture is there shown such a sense of
childlike dependence upon the Giver of all good (including all evil), or such faith in the unfailing mercy
and free forgiveness of God as in the Psalms and in
the Prophets.

The Hebrew thought of God was the child's thought
—the child's intimate thought, and had in it a naïve

feeling not discoverable in the early pagan thought. The latter was more completely crystallised in its expression, more definitely projected in the form of myths that sought correlation and consistency, while the Hebrew thought became neither mythology nor theology, being withheld in that flowing realm where all life is a constant miracle—a field of easy transformations, of shadowy appearances that come and go as in a dream, of living truths completed in their own contradiction.

The instability of his environment impressed the Hebrew. Existence seemed to him like the fluidity of water, now lifted up in unsubstantial vapour and again taking visible shape, falling to the earth and dispersed over its surface—a blessing even in its descent and dispersion. Formal ethics was as impossible to him as was fixed dogma. It never occurred to him to determine the consistent structure of human character any more than it would to make a chart of the divine nature and attributes, limiting his God by definition. His hope was not a logical expectation; and therefore we do not find in the Prophets any formal determination of Messiahship, which nevertheless we, looking back, can see dominant in living imagination and pregnant phrase. Some issue, it was felt, there must be of deliverance; but when we read in Isaiah of "a sword bathed in heaven" we know better than he what depth of meaning was in his words. There is no generalisation in the expression of the great hope; the imagination always takes a concrete shape, but capable of expansion into what we see is the spiritual principle of a new kingdom, as when the prophet foresees some reconciliation to come of good and evil : "The wolf also shall dwell with the lamb, and the leopard shall lie down with the kid ; and the calf and the young lion and the fatling together ; and a little child shall lead them."

For the Hebrew there was no logical plan of any life. He saw no anomaly in the suffering of the innocent for the guilty—the procession of life was in no other way; it was a course of vicarious passion from generation to generation. In the time immediately preceding the coming of Christ, the belief gained ground and became a conviction that the sufferings of an innocent man were of living value to the race, having not merit as satisfying divine justice, but a communicable virtue in the action and reaction of a life wherein formal justice had no place. Reaction, often taking the extreme form of contradiction, was so familiar to his experience that the Hebrew conceived it as prominent in divine as in human operation. The wheel was always turning, so that the low were lifted up and the exalted were cast down. His beatitudes were, like those of our Lord, apparent paradoxes. Repentance, associated in his mind with abject misery, as in the mind of the Prodigal Son, was the great reaction in the life of a man, bringing him home ; and he would as soon have had a god of wood or stone as one who did not himself repent, so that the Father's mood could respond to that of His returning child. To him God was not the Immutable. The visible universe was but His vesture, to be folded up like a garment in His own good time—forever, indeed, being folded and unfolded. What, then, was there which man could wrap

about himself, whether of goodness or badness, that
must not fall away, leaving him naked in the day of the
Lord?   The outwardly built-up character, of whatever
sort, must be consumed in His fire.   The flame which
destroys is the flame of Love; though seeming to
angrily swell and roar, devouring every dry thing,
yet the beginning thereof is the tender yearning, and
the issue the tender renewal of the eternal kinship.
The wheel of life, showing red and black beneath,
shows green in the softness of warmth and light above;
and when the revolving spheres in heaven themselves
grow old, the fire that consumes them destroys utterly,
that there may be completeness of annihilation and so
entire transformation—a new wheeling and sphering of
morning stars.

The flame of life is tropical, forever turning and finally rending.   But the obverse of nothingness is Creation.   Therefore the Hebrew, in the loftiest strain of
his spiritual imagination, loved to dwell upon the signs
of destruction.   To him life presented the constant alternation of wrath and love, of storm and peace, of dark
oblivion and softly rising dawns of remembrance.   In
the extremity of affliction he rent his clothes; then he
anointed his head and washed his face.   This man of
sorrows was anointed with the oil of gladness above
his fellows.

XIII

The anthropomorphism of the Hebrew was implied
in a faith which so closely united the divine with the
human. The earliest conception of this union was that

of a flesh and blood kinship, and sacrifice in its original form was a feast, celebrating and renewing this intimate covenant. The Hebrew
continuation of this old Bedouin kinship,
while it was really a transformation, yet maintained the intimacy of the divine relationship. Blood
was still its living current, and, wherever shed, returned
to its source. Next to the current of life, the increase
thereof, whose symbol was fatness, was held especially
sacred, and in sacrificial rites the fat which was burned
went up as a sweet savour of grateful return for the
abounding of life, as the blood shed was a response for
life itself. The Hebrews were forbidden to eat the
blood or the fat of an animal, since these are the
Lord's. Perhaps it was from this association that
Swedenborg regarded fat as the celestial principle.
With the Hebrews it was associated with the feeling of
mercy and compassion, and was the sign of bounty;
but its essential mystical significance relates to abundance not as plenty, but as increment—the power of increase which is so pre-eminently the miracle of life in
its wondrous fertility and growth. In the spiritual as
in the physical world the first of all commandments is
" Be fruitful and multiply." Herein also is the principle of authority (*auctoritas* from *augeo, to increase*), the
gracious marrow of our hard bones. The perversion of
the principle is avarice, oppression, hardness of heart,
scripturally indicated in the phrase, as applied to a
man thus degenerate, designating him as " enclosed in
his own fat." In the Oriental conception the beauty
of woman was the favour of *embonpoint;* and according
to the most recent deliverance of embryological science

(margin note:) Hebrew Symbolism touching Incarnation.

the better nourished ovum becomes the female. In maternity the two sacred Hebrew symbols are united. The blood of the mother is turned into milk, and from the roundness of her breasts flows into the roundness of cheek and limbs that give to infancy its grace and favour.

Attention has already been drawn to the fact that Hebrew symbolism was confined to a living, growing organism, as distinguished from æsthetic re-presentation *in alia materia*—in stone or on the canvas. We see in this symbolism, as above indicated, an especial confinement to a human body, as the real spiritual temple —"the temple of God" in St. Paul's interpretation. It is the carnal which becomes the Incarnate : cast down to hell and lifted up to heaven. The symbolism reached its most profound meaning in the words of Christ : "Except ye eat the flesh of the son of man and drink his blood, ye have no life in you." And he who said this, in the same breath repudiated the flesh as profiting nothing. "The words that I speak unto you—they are the spirit and they are life." While this is a contradiction of the one declaration to the other, both together are really an expression of the identity of embodiment with spirit, "He who hath seen me hath seen the Father." The vine which has been so long tended and pruned has come to its fruitage ; its grapes have been trodden in the wine-press, and here is expressed its free spirit—that which was its life from the beginning.

The Hebrew idea of spirit implied personality ; it was not an abstraction. Therefore, the adjective "spiritual" was not in use. It occurs but once in the Old Testament (Hosea ix. 7), where it has not the modern sense, and never in the Gospels, though so frequent in St. Paul's Epistles. The phrase "spiritual life," so familiar to modern thought, is not to be found in the Bible. The spirit must have embodiment, and could not otherwise be conceived. Thus the Spirit of God descended upon Christ when he was baptised, taking the body of a dove. To the polytheistic Aryan this Spirit would have taken diverse shapes in numberless divinities — dryads and naiads and nymphs ; but to the Hebrew it was the One. Pagan divinities were given the human shape ; in the Hebrew faith man was fashioned in the image of God, and though the visage of humanity was marred, yet had the Divine Spirit seized upon the seed of Abraham for the renewal of His image. The divine kinship was to be realised in the flesh, and in a sense far deeper and more intimate than that in which the Jews "had Abraham for their father." This intimacy is sometimes expressed in the scriptural phrase, designating a man in a state of peculiar exaltation as "in the Spirit." They were sons of God to whom the Word came, and, in an especial sense, Christ, who was the Word become flesh.

In Christ the Spirit, which had been veiled and hidden, was revealed as free—in a mystery openly wrought in his very body. " I have power to lay down my life, and I have power to take it again." Is not this the full expression of a freedom which may well be called that " of the sons of God "—the breaking of a circle hitherto closed as to human vision it seemed, or, rather, the completion of the circle by showing, in the Resurrection, the other half of it, hitherto shadowed by the apparent conclusion of Death ?

XIV

We must be on our guard against the conception of
what we have called the Hebrew destiny, as being, because it was so singular, something contrary to the
course of Nature, when that course is truly seen.   The
The Singu- parabola described by a comet seems singularity not lar to the denizens of planets moving in ellip-
Supernatural. tical orbits, but we do not therefore exclude
this phenomenon from our science of astronomy.   What
we call supernatural, applying the term to any singular
manifestation of life, is something in nature itself which
is inexplicable through any co-ordination we have been
able to make.   Even the mystical view, which transcends the visible in its intuition of creative life, only
postulates the hidden side of nature—the fountain of
its issue ; as if, recognising the visible as development
in form and structure, and in a harmony imperfectly
comprehended by us, we saw also, with the poet, that
" All foundations are laid in heaven."   While we are
naturally apt to think of vital systems as planned, all
forms having been divinely premeditated and all relations preconceived with reference to adaptation, still
we know that the creative must *be* the formative and
involve the adaptation, and that the admission of a
single arbitrary element, such as we associate with
human design and the adaptation of means to ends,
would introduce into the universe the operation of a
limited wisdom—not of wisdom spontaneously coming
under a limit, but finite at its source, and liable to the
fallibility and uncertainty attending all human experi-

mentation. Moreover, we know that even in human life—in all that determines its real issues, as distinguished from ends consciously in view—there is no such arbitrament, but rather a vital destination from a purpose that cannot fail, inerrantly wise.

The Hebrew was no more a man of destiny than was the Assyrian, the Chinese, or the Indo-European. In the physiology of humanity, each of these races had its special allotment of function by a vital destination like that which determines the drift of constellations, the configuration of continents and the currents of the air and the sea. Yet the mission of the Hebrew was as peculiar and distinct as are the course and temperature of the Gulf Stream in the midst of the waters. It cannot, in our thought of it, be separated from the incarnate Lord, to whom was given "power over all flesh," so that the mystery of the Incarnation, though so intimately associated with the seed of Abraham, is yet catholic and genetically dominant as associated with the destiny of the whole human race.

XV

The repudiation of Christ by the Hebrews is as remarkable as his acceptance by the Gentiles. "He came to his own, and his own received him not." This was but the continuation of the hostility shown to the Prophets, of the recalcitrance of this obstinate and stiff-necked people against its peculiar destiny from the beginning. There had been the deepening of a vast hun-

Attitude of Jew and Gentile toward Christ.

17

ger — in the few, indeed, for the bread from heaven,
but in the many, especially at Jerusalem, for earthly
rehabilitation. To these latter the Son of David, the
long-expected redeemer of his people, seemed only to
aggravate their desolation. He despised the glory
upon which their hearts were set, bringing their pride
to the dust even as had the Prophets before him. He
repudiated them, even their boasted kinship with Abraham and their consciousness of an especial divine elec-
tion ; his sermons and parables exalted other peoples
at the Jews' expense ; he predicted the destruction of
their temple. He put aside his own mother and brethren in favour of a more blessed kinship. He chose for
his companions those whom the pious Pharisees and
Levites deemed outcasts. He crucified them, and they
crucified him. He told them that publicans and har-
lots entered the kingdom before them ; they preferred
Barabbas, the robber, to him, and condemned him to
die between two thieves. He seemed to Judas false
to a cherished hope, and Judas betrayed him. Even
his friends, those who believed in him, were compelled to drink the cup of bitter humiliation at his
defeat and death, and to listen to the jeers of them
that said in scorn : Others he saved, himself he could
not save. In this dark hour his disciples were stricken
with shame and fear, and one of them denied him.
Failure was turned into triumph by the Lord's resurrection ; but the full meaning of this glorious morning
was not appreciated by the believers who remained at
Jerusalem, clinging to the old ritual and still rejecting
the uncircumcised, while the sect of Ebionites wholly
misconceived the new life, ignoring its positive princi-

ple, which was to revitalise and transform the world, and continued beyond the Jordan the practice of a sterile asceticism, maintaining that divestiture which in itself was merely a negative and accidental aspect of the Christ-life.

The hunger of the Gentile for the Christ was due to inanition, to the vanity of earthly accomplishment, and was a downright malady; like the fever of the prodigal, who, having been sated with revels, had been brought to starvation, while the Hebrew, like the elder brother in the parable, had been kept, albeit by a kind of compulsion, in the Father's house. The Gentile had come into a barrenness which, left to itself, must become utter sterility, as of a rod that could not blossom. He had not been tormented by the consuming flame of a sacred fire or by a school of prophets forever cutting his life back to its root. His oracles were dumb; his temples, adorned with the statues of divinities, were haunted by the ghosts of dead gods and not filled by a living presence; his ritual was more easily repudiated than that of Jerusalem could be by Hebrews as devout as the apostle James. Therefore he not only with greater avidity accepted the new faith, but was more alive to its newness, and readier to give it a practical embodiment, making our christendom.

Very likely, if we had the means of ascertaining the historical truth of the matter, we should find that among the Jews those who most eagerly embraced Christianity were the Pharisees, not only because the idea of the resurrection associated with the early and beautiful faith of this sect had been in so remarkable a way revived, but because their religious observances

—like those of the pagans—had become so formal in minute and trivial details as to the more readily fall into oblivion and disuse. Certainly, after the resurrection of Christ, we have no record of their opposition to the new faith, and Paul, who boasted himself "a Pharisee of the Pharisees," was the great apostle to the Gentiles.

The Gentiles, who stood outside of that earthly kinship which related the Hebrews to Christ, and who had directly no part in his birth, have yet given him his embodiment in the world, overshadowed by the Holy Spirit for a new conception of the Emmanuel. Therefore said Isaiah:

"Sing, O barren, thou that didst not bear; break forth into singing, and cry aloud, thou that didst not travail with child: for more are the children of the desolate than the children of the married wife, saith the Lord."

Another Sarah laughed in her tent. Another virgin was to magnify the Lord.

XVI

The real meaning of a movement is disclosed in the issue. The personality of Jesus was the issue of the Hebrew destiny. He was the Child of that race, and as it is that which is to come that is dominant, his singularity determined the singular character of his people from the beginning, who thus became the progressive incarnation of him. In him was concluded this embodiment through a flesh-

The Universal Hope.

and-blood kinship. Through that more intimate kinship with the Father, which it was especially his mission to reveal and make real for all men, Christian humanity is a new incarnation of him, as in a spiritual
body. Thus the Lord is ever to come, reappearing in
every renascence of human society. The Divine name
Jehovah, or *Javeh*, as Dr. John De Witt has shown
in his version of the Psalms, means not merely *I am*,
but *I am to come;* so that in the largest sense all manifestation is his appearance. The history of humanity
is a divine history.

What we commonly call history is a record of structural development ending in decay. In a spiritual interpretation of human history we see death only as
birth, regarding not merely what falls but also and
chiefly resurrections — a line of successive manifestations ever newly revealing the Father; and in such a
view we trace from the earliest record to the present
time a more or less distinct line of progressive revelation. It is a prophetic line, as remote as possible from
any sacerdotal association, yet ecclesiastical in the original meaning of the term (from *ecclesia*, a calling out)
since the call of Abraham out from his country and his
people into a new promise and possession. It begins
at every epoch, like all new life, in dissociation and repulsion, disclosing in its development the bond of attraction and association.

This line is human before it is Hebrew. To us Abraham appears as the first of the prophets, but in some
more primitive faith what line may have preceded him
of men who heard the divine voice calling them out
from among peoples degenerate in custom to begin in

another land a new order, conserving a seed of promise
for mankind? Who knows what nursery this earlier
church may have had, perhaps in old Accadia, from
which comes to us a faint breathing of the eternal
hope? The figure of Melchisedec stands boldly out
against that ancient sunrise. And, before all, was not
the promise made to Abraham first made to Eve, so
that divinity was bound up with our very mortality,
seizing upon "the seed of the woman" in the beginning of generations? In the Gospel of John all the
generations of time bear the impress of this hope, and
we behold the Logos as the light of the world, the
glory of an Evangel coeternal with God.

In the identification, from Eternity, of man with the
Lord is held, behind all veils, the living meaning of the
Universe.

XVII

But it is with Abraham that modern history begins—
our history, the warp and woof of whose variegated
web may with more or less certainty be traced to its
original patterns. This venerable patriarch;
the friend of God; the father of many peoples besides the Hebrew; the peace-loving
brother; owner of flocks and herds and gold and silver;
the victorious warrior honoured of Melchisedec; the earliest of Semitic sojourners in Egypt; the first merchant
on record dealing with money; the zealous intercessor
with God for the doomed cities of Sodom and Gomor-
rah; and the father of the faithful, whose Paradise was
his bosom, represented to his descendants the golden

The Setting of the Type.

age. David and Solomon were glorious memories to the Hebrews, but the thought of Abraham carried them back beyond their trials and distresses to a period of calm content associated with spiritual promise, but not with the fiery furnace through which they passed to its fulfilment.

The long patriarchate was, as has already been indicated, a happy preparation for the peculiar life of the children of Israel. Its background stretched far back into the Bedouin past. The deep impulse which sent Abraham forth from Chaldea, instead of disturbing the patriarchal habit of tent and shepherd life, gave it distinct form and character. Thus was nourished the genius of the race, and doubtless if we could penetrate the veil which hides from us all but the superficial aspects of life in those early days, we would be able to note even there the singularity afterward so conspicuous, and, in the dreams of the shepherds as they watched their flocks by night, discern some tokens of a mood not elsewhere deepening and expanding, but there alone increasing, inbreathing, and infolding God, and making for Him spacious reception, and so enlarging the capacity for the spiritual promise—for its heavenly hope and its earthly desolation. It was a mood prescient of the Psalmist who should sing. "The Lord is my shepherd;" especially was it prescient of the words of Isaiah: " Enlarge the place of thy tent, and let them stretch forth the curtains of thine habitations; spare not: lengthen thy cords, and strengthen thy stakes."

Time was given for this enlargement, for the expansive culture of the shepherd's dream, full of the night

and the stars and God; but it was of eternity rather
than of time, so that the mood of it was a strong holding of things inwardly precious and incorruptible, and
a strong withholding from artificial constructions—from
the things which make cities and kingdoms and the
institutions of civilisation.   The religious instinct common to all peoples was in these tribes lifted out of its
usual plane of development.

From the first, then, the singular type was set, which,
though it had so little outward stability, was, as it always has been and is to-day, the most insistent and
abiding racial type on earth.   Even in the primitive
patriarchal era there was something more than a noble
quality of animal life, than the strong instincts of a
vital manhood, fierce in its virility, yet with a natural restraint; all this was exalted and intensified by
that divine alliance which was already recognised as
a reality, the ground of the later covenant, embracing a world.   Some special readiness to receive was
the basis of a special revelation, though the reception
was in trembling fear and with many signs of repulsion.   They were themselves gods unto whom the
word of the Lord came, else it could not have come;
some consubstantial flame in man was witness to the
flame of the spirit.   That which was in the heart of the
child Samuel — that waiting desire which made him
listen in the still night for the divine Voice—that which
in the inmost heart of man makes it the bride of God,
was a determining element in the vital destination of
the Hebrew.

The nomadic shepherd life had always some unrest.
The tent was forever being shifted, if only for new past-

urage ; but there was in this wandering impulse, as affecting the early Hebrew, a spiritual disturbance unsettling content, the expedition of a mystical pilgrimage.

As in all childhood there is a heavenly holding and withholding, which in some one child becomes a special nurture with more ample storage of buoyant hope, a deeper inbreathing of the air of dawn, so, while we discern in all race-beginnings a spiritual impulse, a fresh and living flame like that which breathed through the Vedic Hymns, yet in the Hebrew origin we behold such seizure upon God that the divine seems to be insphered in the human, increasing and abounding there through the long morning ; and, though that which holds it so largely is finally broken, it is broken as is some precious argosy whose treasure is bestowed upon all lands—is indeed the broken matrix which has held Emmanuel.

In the prolonged patriarchate was set the type of this peculiar people—the note to which it was held accordant, though the discords were many and violent. Here were engendered Psalm and Prophecy and the Messianic hope. This period lasted long enough to become an exemplar. The tent, so easily folded and removed, was the foretype of that earthly instability which characterised the fortunes of this people—an ideal standard of divestiture to which the prophet was always calling back, reducing life to its simplest principle. The familiar watchword, "To your tents, O Israel!" remains to the last the refrain of Hebrew history.

How different this Hebrew patriarchate from that

of the Chinese, which we see to-day in the crystalline
calm where it has been held in arrest for centuries!
How different from the obese degeneration of the ancestral type among the Bedouins of the desert! And as
to its outcome, we see clearly how distinct it is by comparing the Hebrew religious movement with that which
emerged in Islam — the difference between enslave-
ment, defeat, and captivity disguising heavenly dominion, and that kind of possession and conquest
which is the dissipation of spiritual energy.

The peace which the Hebrew loved, the longing for
which led him inland while the adventurous Phœnician
sought the mastery of the sea—that rest besought by
the Psalmist, such as the dove seeks in its flight : these
stand out in pathetic contrast against a troubled career
of fiery trial and chastisement. It is just such a contrast that impresses us in the personal life of Jesus,
between the serenity of Galilee—that charmed circle
of security from which he sends forth his defiance to
Herod—and the fretful tumult, the cruel hostility of
Jerusalem. The deepening of capacity is for the larger
inclusion of pain and strife, as well as for that of a
heavenly peace ; and so it was in the divine life of the
Son of Man, who had not where to lay his head, who
took the stings and arrows of every enmity, and who
not merely suffered evil and death but included all
evil and all death, so that his rising again might stand
against all falling. He descended into hell, so enlarging the scope of that descent that it emerged in heaven.
Before him, neither in pagan nor Jewish thought, was
such emergence conceived as possible, just as before
him the mortal issue was not seen as life.

The idea of heaven as the eternal habitation of souls freed from earthly bondage is so familiar to us that we are apt to forget that it is wholly a creation of the Christ-life and the Christ-death, The Opening of Heaven. followed by his resurrection and ascension. The phrase "going to heaven" is strictly modern, and as indicating the direct destination of a departed spirit is quite wholly Protestant, since the great majority of Christians believe that there is an intermediate state. To the martyrs, as to Stephen, heaven seemed to open for their immediate reception ; but to the Lord himself there was no such invitation. Therefore he said to the penitent thief : " This day shalt thou be with me in Paradise "—meaning that happier part of Sheol allotted to the faithful. Hitherto, in the hope of pagan or of Jew, the movement of the soul had been arrested at this point, as if by fixed conclusion. The only Hebrews in heaven were those who had been divinely translated thither—Enoch and Elijah, and, it was believed, Moses ; besides these, it was the abode of God only and the angels.

The Lord never directly promised his disciples entrance to heaven, though intimating that in his Father's house were many mansions, and that he would prepare a place for them—praying, moreover, that where he was they might be also. The resurrection in which the Pharisees believed was a return to earthly embodiment and habitation. Only the Lord's ascension opened heaven. The Gentile Christians, by swift re-

action, readily accepted the idea of the supreme exalta-
tion ; but the Hebrew, as shown in the Apocalypse of
St. John, expected the descent upon the earth of a new
Jerusalem.

The idea of place, in this connection, has no importance; what is really significant is the ascension, as the
complement of so deep descent—the escape from that
old and sterile conclusion in Hades which had so long
impressed the minds of men as something inevitable—
the completion in ineffable light of the soul's wandering that hitherto seemed to have been arrested in darkness. The descent was not evaded ; death still awaited
every man, and the grave deepened into the Inferno,
but the cycle was completed, and what had been
bounden was free—the bond itself finally shown as a
home-bringing of God's children to the bosom of
another Father than Abraham.

It is interesting to trace the adumbration of this
freedom in the Hebrew consciousness. The primitive
thought of another world was backward and downward,
but with the spirit of prophecy there was a turning of
the face to the light, forward-looking. After the Captivity the idea of angelic beings became more and
more familiar to the Hebrew, mingling with his hope
of resurrection, though the angels should descend to
him rather than he should ascend to their abode. The
Lord spoke of the children of the resurrection as becoming like the angels in heaven. From this it was
only a step to heaven itself—but that step halted. Then
there was that last week in Jerusalem, with its gathering trouble, relieved by visits to the restful home of
Mary and Martha in Bethany; the raising of Lazarus,

and the evident expansion of some mighty and luminous thought in the mind of Jesus, prophetic, absorbing, withholding itself from expression even to his disciples, as something they could not yet bear and which must
await disclosure from the spirit—from that free spirit
which was in him, made wholly free when he should
"go away." Flesh and blood could not reveal it, but
rather the vanishing of these. With the resurrection
of the Lord—which, though it only brought him back to
the light of earthly day, still seemed to remove him from
the accustomed familiarity, so that he only at times
suddenly appeared to them for brief converse and then
as suddenly vanished—their spiritual sense was deepened. Their hearts burned within them while he talked
with them on the way to Emmaus, showing them what
was the real meaning of his sufferings and death and
resurrection, as completing the divine mission of Israel
in the person of the Messiah. "Ought not Christ to
have suffered these things and to enter into his glory?"

The consummation of the lifting power of the life
manifested in the Christ was reached in his ascension.
He who had "descended into the lower parts of the
earth . . . ascended up far above all heavens, that he
might fill all things." The tree of Life could not fill
the heavens till its roots had taken hold of the nether-
most abyss.

Therefore it is that, for the Christian, Death and
Evil are deepened to the utmost, and in like manner
the consciousness of Guilt, that nothing may be left
outside of the comprehension of the lifting Life—that
the ascent may "lead captivity captive."

XIX

The Hebrew movement, thus consummated in the Christ-life, represents the epos of the human soul, not in such terms as the ancient poets used in their epics celebrating heroic adventure—the quest of the Golden Fleece or the taking of Troy— but far withdrawn from any idea of mere outward accomplishment and confined within the scope of a spiritual destiny expressed in terms of living guilt and living righteousness.

The Hebrew Idea of Sin.

Childhood is unmoral. It has the primary conscience, whose instinctive feeling is not expressed in abstractions or in such judgments as are ethical in our modern sense. It has natural control, a vital restraint, deeper and surer than that which is concerned with external relations and consequences rationally considered. The Hebrew, keeping much of the plasticity of childhood, had this living conscience, not merely in the sense in which all primitive tribes have it, but in that sense exalted, so that sin was felt to be blood-guiltiness, as violence of the bond of kinship between men and God. When the Prophet wished to convince David of his great sin, he did not refer to the broken law but to the home he had broken. So the Lord made the test of a divine judgment not any dogmatic or ethical condition, but only tenderness of heart toward all men as toward brethren; as if this were itself the fulfilment of the law. He who was to come with the spirit and power of Elias was to turn the hearts of the fathers to the children, and of the children to the fathers: it was

the concern of kinship. Paul's definition of " religion pure and undefiled " points to the same living truth.
Not moral perfection but newness of heart is the vital
distinction : the newness is for tenderness. The idea
of sin entertained by the Greek and Roman was confined to failure, with reference to that outward completeness which was to them the chief end of life.
The Greek word for sin means the falling short of a
mark : some outward standard is implied. The older
idea, expressed in the Latin *ne-fas*, was originally allied to
the Hebrew sense of guilt ; but this meaning had been
outgrown, surviving only in the lingering regard for the
Lares and Penates, the deities of the hearth, and in that
tenderness of piety which never became wholly extinct,
and which, indeed, was the great softness that, turning
into manly fibre, was the basis of Roman virtue and
mastery. Rome made for herself a world of dependent children by somewhat the same quality as that
whereby, maintained in its plasticity, the Hebrew became a Child for the world.

XX

The Hebrew movement, culminating in the Christ,
was a discrete destiny, necessary, once and 'for all, to a
singular issue — to the Appearing, in time and in the
world and in human form, of Eternal Light and Love—
an Appearing so wonderful that we ask how
it could have been, and yet so longed for,   The Issue.
and so resuming all other appearances in Nature and
humanity, as their central illumination and essential

glory, that we ask how it could not have been! God
so loved the world: the world so desired God! Because the sun is in the heavens the waters that run into
the sea are lifted again to their native heights; and
so, in all ways, is the pulsation of the physical earth
maintained.　How else could there be the full pulsation
of the spiritual world save as its sun responded to the
desire in the heart of man?　Christ is that Sun.　We
were in him, though we knew it not, and he appeared
in us and to us.　He descended and he arose, and he
stood for our falling and rising, and we saw in him
what we turn from, as worlds from their light and that to
which, following the same old planetary habit, we forever return—what we deny and what we confess.　Apart
from this movement, which had for its issue the Eternal
Child, full of grace and truth, shaping for us the language of a new kingdom, we should be at a loss, having
no clew to our labyrinth leading outward into freedom,
no escape from our entanglement.　That which is hidden could never have come to the light.

To suppose this movement as not having been would
be to suppose humanity—the ultimate specialisation of
cosmic life—to be completely insulated, an island from
which its embosoming ocean could at no point be seen.
The spiritual loss might be compared to the sterile
physical existence of man upon the earth, supposing
human life to have no hidden fountain in its organic
cell structure whence proceeds any child.　The newness, of which the child is the symbol, is the charm of
existence, the charm of an expansion renewed by attraction, of desire renewed by death.　As in the rising
again of the Lord—the " one sign given unto men "—the

way of death was seen to be the way of life, this resurgence, to the early Christians, stood for a new childhood; it was the transcendent Nativity, whereby they
were "the children of the Resurrection."

CHAPTER II

THE PAULINE INTERPRETATION

IT has been charged against Christianity that it looks ever toward a dying Lord, drawing always near to the grave, emphasising sin, also, as it does mortality, and clothing itself in a sorrowful habit, loving rather to dwell in the house of mourning than in that of feasting. This attitude has been contrasted with that of pagan philosophy, which appealed to aspiration and extolled virtue, finding the highest excellence in outward accomplishment and inward serenity.

As in no ancient faith was there the exaltation of a sure and steadfast hope such as lifted the heart of Israel, so was there never such a sunburst of dawn as that which exalted and illumined the hearts of the early Christians. Nevertheless it is true that these Christians turned their faces away from the vision of any earthly sunrise, literally as well as figuratively faring westward, renouncing the hallowed traditions and associations of the Holy Land, seeking discomfort, courting persecution, facing death in every Roman amphitheatre, and leaving upon their tombs the only inscriptions of their faith recoverable from this period of their tribulation.

The Apostles were the witnesses to an eternal verity, disclosed in their Lord's resurrection—that death is in-

deed the unseen angel of life, with wings that lifted to
heights beyond the reach of mortal vision and earthly
aspiration. Death had not befallen the Lord, but he
had pursued death, had clothed himself in the mortal
habit, and in its corruptible had shown its incorruptible.
The followers of Christ, therefore, sought not safety;
their pilgrimage was not away from the City of Destruction but through its flaming streets. To them, indeed,
every city, every structure which had been raised by
human effort seemed about to fall. They were on fire
within, and imagined a world on the verge of conflagration; the framework of Nature as of all human systems
seemed "like an unsubstantial pageant" soon to disappear, dissolved in fervent heat. They built no church
edifices and established no elaborately formal rites.
They took no active part in the social or political functions of the world about them.

St. John's Apocalypse and St. Paul's Epistles disclose
in different ways the prevailing conviction that the end
of things was at hand. In John's vision everything
seemed to vanish before the "wrath of the Lamb."
Paul looked for the speedy emancipation of a universe.
John saw a new Jerusalem, Paul "a new creature."
The Lord had said that the Gospel would be preached
to all nations before the end of the world, and to them
the swiftness and magnitude of the pentecostal revival
seemed the beginning of a movement which would not
halt short of its rapid consummation. St. James alone,
with steadfast zeal for the ritual of his fathers, was
conservative and temperate in his expectation, representing ecclesiastical stability, and probably for this
reason he gave more consideration to the ethical side

of a Christian life, emphasising the value of good works.

Each of these apostles was mistaken in his forecast of the immediate future, though each, through the distinct phase of his hope, contributed to the completeness of that testimony which is known to us as the "new testament." No greater confusion fell upon Peter when he reflected upon his denial of the Master than awaited James when he was overtaken by the destruction of Jerusalem ; and we can imagine the consternation of John if he could have foreseen the position which was to be held in Christendom by that Rome upon which he saw emptied the vials of God's wrath, or that of Paul if he could have followed the lines of Christian development into an ecclesiasticism more elaborate than that of Jerusalem, and have seen the world which he felt crumbling beneath his feet enter upon an era of unprecedented stability in every field of human activity.

Paul's thought never crystallised into either a philosophic or theological system ; it was so close to a nascent and flaming life that it was luminous with its light and plastic to its creative spirit—a quickening spirit that impelled his swift journeyings over the whole known world even to its westernmost limits, and at the same time gave him the deepest insight into the mysteries of the Christian faith.

It was because Paul's faith was fixed upon the invisible and the eternal that the whole visible universe seemed to him so unstable. "The fashion of this world passeth away." This dissolving view was ever present to his mind, because he felt the power of a new creation. He

dwelt upon death because his chief theme was the resurrection, and upon sin, which is the sting of death, because for him had arisen the Sun of Righteousness. The
principle of a new life dominated his thought. It was
not a new life as having just begun to be. The power
which raised Christ from the dead was the creative
power from the beginning, hidden under the masque of
Nature's bondage—hidden also in the heart of man. It
was the power manifest in the world as a vital pre-
destination, not to be thwarted by human traditions or
aims ; the power working in evil as well as in good, in
the hardness of Pharaoh's heart as in the faith of Moses;
the power of the law as well as the power of grace.

In the light of the renascent spiritual principle, Paul
saw a new humanity, as from a second Adam, and a new
creation. A revelation had been made, which gave a
new scope to human life, and a new meaning to the
universe ; but it was the first purpose of the divine will
for man and the world, though last in the express brightness of its manifestation. It was not a new cycle of
human or cosmic life, but the completion of the old :
the fulfilment of its divine meaning. The natural body
was raised a spiritual body, and so the natural man was
raised a spiritual man, growing into the stature of the perfect man in Christ. The living soul, animating the flesh
and boasting in the works of the flesh, was disclosed as
the quickening spirit, one with the Father, and inspiring
a universal fellowship, which had been from everlasting
but was now for the first time luminously real to human
faith.

Paul regarded human destiny as inseparably bound
up with that of the universe. The visible world, though

imaging the spiritual, "was made subject to vanity," under the bondage of corruption, and man, as a part of this creation, was under the same bondage, and therefore mortal and sinful, the sting of death being sin and the strength of sin the law. The subjection was a divine limitation and was universal, pertaining to all visible manifestation : it was not of the will of man but of God. As the whole order, seen in its natural operation in time, was clothed upon with the mortal habit, every structure being brought to naught and thus "subject to vanity," so in humanity there was a special death and a special evil. Sin was something more than was definable in particular acts : it was a state. Thus Paul speaks of Christ, who knew no sin, as becoming Sin, wholly identified with man in his limitation. Paul emphasises the descent of Christ, thus bringing him into the estate of falling man. As Christ had all of death, in its essential meaning, and yet saw not its corruption, so he took the inmost reality of sin in such wise that the divinity of it was blamelessly transparent in its humanity—its crimson ever turning white, as was natural in a life essentially redemptive, and whose blood flowed for remission. He was the reconcilement of sin with the eternal life.

Thus the bondage itself came to be seen as the bond of kinship ; showing that originally it was this in its divine ordinance.

Paul contrasts, by sharp antithesis, the works of the law with the operation of grace. But the law was, at its fountain, holy—a fatherly commandment, a gracious provision suited to fallible humanity, and even with its thorns hedging in the truant nature, pricking the conscience, convincing of sin. While it hardened with the hardness

of the human heart, and the men who sat in Moses'
seat laid upon the people burdens too grievous to be
borne; while it became itself a part of the bondage,
yielding to corruption, so that the works of the law partook of the vanity of all outward accomplishment; yet
this very inanition was a preparation for the gospel of
grace. Thus the law was a schoolmaster leading men
to Christ.

To Paul's vision was opened a spring-time for the
whole world, with issues unforeseen, indeed, and immeasurable, but whose meaning had been fully disclosed. A new principle, hitherto hidden beneath the
mortal masque, was manifest. The followers of Christ
need not turn away from death, or regret any outward
desolation, however complete the divestiture. Death
could not bankrupt life, being indeed its only solvency;
though it stripped the soul of its investment of good
works as of all other vesture, the nakedness was that of
the child of the kingdom of grace; this absolution
took no note of works of merit any more than of any
other works. This death, moreover, unmasquing all else,
put aside also its own disguise, repudiating mortality.
But for this absolute newness there could be no deliverance from the body of death.

So significant was the resurrection of Christ to Paul :
the revelation of a new death hidden in the old, even
as the spiritual principle is hidden in the visible world.
What was implicit in the bondage had become explicit,
a manifest redemption. A body inviolable and incorruptible had been returned from the grave, raised by a
power which lifted it out of the closed circle of mortal change and progression in which the visible world

seemed locked — even lifted it up into heaven. The
suspense was broken. This revelation had been made
not in an analogue, or symbol appealing to the mind,
but in an appearance to the sense—like a flash of the
Eternal into Time — of a spiritual body. Thus was
shown the fashion of the world to come, into which all
vanishing things were transformed, so that the universality of death was the hope of the universe.

It was not merely the illumination of a truth, but
became a living, working principle. In its light the
Christian could not only face death, but anticipate it
by the inclusion of it in life, and thus bring into earthly manifestation the power of the resurrection, lifting up the spiritual man just as the nutrition and
function of the physical man were from the inclusion
of death for the resurgence and ascension of the organism.

This anticipation of death was the essential condition of a new life, in a Christian fellowship, on earth.
If the Kingdom of God was to have an earthly realisation, it must be through dying daily, both for ministering and to be ministered unto. To become like
children, after this new type of childhood, meant a
withdrawal from the world for spiritual expansion at
the same time that contacts with that world were multiplied—channels for freely receiving and then for freely giving. Baptism was a burial with Christ that the
Christian might rise again with him, lifted up by the
same spirit. The burial and the rising, begun in this
symbolic rite, were repeated continually in the pulsation of Christian life—the vanishing side of which was
a hiding with Christ, the return beat spiritually vital-

ising the outward body of the individual and social organism. On the one side the flesh was denied, on the other it was made the temple of God, and upon the heart of flesh was freshly inscribed the law in its original terms of love.

The formal obedience of the law in every point could not in itself secure deliverance from its bondage; might, indeed, result in self-complacency and pharisaical self-justification, the very habit of such obedience becoming automatic routine, and ending in a stoical acceptance of death. In this fulfilment the law destroyed itself, became crystallised in a heart of stone, losing its proper virtue in a life thus arrested, and death would be liberation only as breaking up the brittle structure and so forcing the final confession of a corruption resisted and denied but inevitable. On the other hand, the willing acceptance of the mortal state, its burdens and its bondage, by the tender hearts of God's children, judging not, repudiating merit, failing at every point, as fail they must, yet having faith in the Father's love, and lifted from every fall by His grace into living righteousness—this obedience is that love which fulfils the law, eclipsing and transcending its letter, and rising into its spirit. Thus are all systems, whatever virtue they may have, urged on to their mortal issue for the regeneration of goodness itself.

The quickness of life, including death as itself a quickness, a lifting and transforming power, was re-creative, making a new or newly visible organisation of humanity in a spiritual body—a fellowship setting up new activities, nutritive and functional. As in the

animal organism the maintenance, through nutrition,
of vital activity depends largely upon nitrogen, the
most inert of all elements (excepting the recently
discovered argon, an equally important constituent of
the air we breathe), so the spiritually organic body
includes for its nutrition the death and inertia of the
human world.   Its catholic kinship includes the out-
cast, the reprobate, the utterly condemned, finding in
the extremity of human wretchedness and sin the point
of penitent return.

It was God's pleasure in humanity that was the
burden of the angels' song announcing to the shepherds the birth of Jesus; and we may without presumption interpret the heavenly voice declaring at
Christ's baptism, "This is my beloved Son, in whom
I am well pleased," as especially emphasising the utterance already made to the shepherds, and as celebrating the new birth of humanity.

We can imagine the change which came over the
spirit of man's dream concerning himself and his
earthly station when the Copernican astronomy disclosed the fact that what had been thought the flat
and inert earth—a condemned world, the degraded
footstool of the universe, the alone dead and motionless, and enclosing death and hell in its secret depths,
as if these were its own peculiar possession—was itself one of the celestial spheres, and so restored to its
heavenly place and motion, no longer excommunicate.
That catholicity which included it in the universal
harmony made also a catholic distribution of such
evil as had seemed its singular portion, and its central fires were seen to be like those which tormented

the bosoms of all the celestial wanderers. In the catholic life was included the catholicity of death.

How much more glorious was the revelation through the Son of Man of the divine spiritual kinship, into whose bond was turned the bondage of every creature! Man, the most fallible of all living beings, and who so accumulated death and evil that he seemed to monopolise corruption and to be inert—"dead in trespasses and sins"—was shown to be, because most lost, the best beloved.

The charm and excellence of this new creation could not be expressed in the terms of physical sensibility, of mental appreciation, or even of ethical motives and restraints. All the sensations possible to the most exquisite bodily organism, heightened by æsthetic and intellectual refinement; the sum of attainable power and virtue: these belonged to a world which had dwindled into insignificance in the presence of a kingdom whose activities were characterised by Paul as "the works of the spirit." The law of altruistic service and sacrifice had belonged to every order of existence, and belonged also to the new, but was distinctive to the latter only through its heavenly transformation and reversion from the measurable merit and value, hitherto associated with its human expression, to the original and immeasurable grace which is the quality of a creative act. The Lord, as the bridegroom of humanity, lifts it into participation with himself in creative action, and in this conjugal relation humanity is no limited saintly company, but a catholic and spontaneous fellowship.

All were under sin, and in all the new creation was

redemptive, "especially in them that believed." While
the tree is known by its fruits, and the early Christians
showed an outward excellence beyond the require-
ments of law and duty, yet those loving believers attached no merit to such performance, and, having
done all, deemed themselves unworthy, and as falling
short of the divine glory revealed to their anointed
eyes. No sum of deeds could fill out the measure of a
life which was supremely Being above all Doing. The
deed could not seem to them other than decrepit, like
the blossom that withers and the fruit that falls. They
tried with glowing lips to tell men what the new creative principle was, but none could understand who
had not their spiritual experience. Paul said it was
Love, and then described every known manifestation of
love as falling short thereof ; for how is one to know
Love save in the wonder of what it is above all it can
do ? It is a creative power, but the creature falls into
impotence.

As a social power, in its primitive manifestation,
Christianity converted altruism into identification. The
neighbor was loved not as another, but as the self of
the lover. Sacrifice was the blending of the human
with the divine will ; not renunciation for mere loss or
divestiture, but for recovery. Suffering was incidentally, and in the sequence of things in time, a discipline,
but in the eternal meaning was one with the divine
passion from the beginning, and as belonging to a living creation. Death and sin were involved in the
resurrection and redemption whereby man became a
new creature.

Paul's idea of dying to sin was not that of ascetic

mortification; he meant thereby a dying to the dying
environment of the old creature and living to that of
the new. Instead of escaping from a sinful world, he
sought every possible contact with it, knowing that
where sin abounded there did grace much more abound.
The seed of the kingdom was sown in corruption.

Thus the Christ-life took possession of the world
with no dainty selection, but, seizing upon the worst,
brought out of it an excellence far exceeding what had
been found in the best. Only a creative and transforming life, drawing its inspiration from a heavenly
source, could have so confidently leaned to all human
moods, following them into their faltering descents, indulgent and compassionate. Thus primitive Christianity followed the pagan world into its own Sacred Mysteries. Paul used the very terms associated with these
for the illustration of spiritual truth. The pagan con-
verts were baptised under a formula conveying the
thought so familiar to them, and so repugnant to the
Hebrew, of a diversified divine manifestation in three
persons.

Paul as the apostle to the Gentiles, as "all things to
all men," turned his face quite away from Jerusalem
and toward the Western world. In his catholicity was
that world fully embraced, while in his doctrine of elec-
tion was expressed the principle of integration, the
principle of the church militant in the development of
Christendom.

CHAPTER III

CHRISTENDOM

I

THE nihilism of the mystic, if cherished for its own sake, would be disintegration, refusing investiture, a sterile simplicity.

The idea that we should attain supreme felicity if we could put aside all veils forever and in a pure spiritual vision always behold God face to face is a dazzling conjecture. Suppose a planet to be able to refuse separation from her sun, would her eternal identifica tion with her lord be any true union—like the "union in partition" which she enjoys in all her varied life? Or, if she might choose, having set out upon her wanderings, to retain her similitude to him—to be forever self-luminous, herself always just another sun, would she not, through lack of contradiction, miss the ultimate dramatic excellence and delight of her destiny?

Espousals.

For, see what happens to the Earth because of her apparent loss and self-desolation. Coming into her hard limitations, she has the inestimable honour of preparing a bridal chamber for the Sun, being nearer and dearer to her lord in her set distance than if she had forever rested in his bosom, for now he rests in hers. As the mother cell is separate from the father cell, so that the latter must go forth, like a hunter to the chase, to possess its sundered mate, so it is in this mystical attraction, first seen as repulsion, which is the charm that binds the Earth to her bridegroom.

Has she pride that she, as it seems to her, is the centre and he the satellite? Or, rather, is it her modesty that she imputes to herself all the inertia and to him all the motion? Really the motion is of neither to the other as to a centre, but both are possessed by the same motion, which is not material but of the spirit. Their union is but the expression of the eternal consubstantiation.

The gain of this planetary bride, the Earth, is through what she has given up. Because of her distance she can be visited by her lord. Divesting herself of her own garment of light, she can be clothed upon with his; hiding her own fires, she can be sensible of his joyous warmth in manifold intimacy. Brought to very barrenness in the diminution of her own force and swiftness, it is given her to sing the virgin's *Magnificat*, and to know that all born of her are the children of the Sun. There is healing in his touch, and all that she perforce distils of poison and bitterness—all the maladies of her desolate nights—yield to his radiant strength. With her he sups and takes up his abode, knowing no delight or charm in the vast distance traversed by his swift wings until he keeps tryst with her. Here only, and not in that blank space, has his face brightness and colour; here only is there for him nutrience and increase and content. This is the garden of his love; of his labour, also, since

here are done his mighty works for his children; and
of his death, since virtue goes out of him with every
revival of earthly life, until he wears the wan smile of
the physician who saves not himself—like the sunset
benediction in the face of Heracles when, after his
grim struggle, he brought Alcestis back to the halls of
Admetus, having himself taken the chill and the mysterious silence.

So is it with all espousals. The union is because of
divulsion, and has the value of distance; its intimacies
have their ground in distinction, which becomes contradiction, like that of a planet to its sun; its special
activities and capacities seek sequestration in a limited
field, "an enclosed garden," sometimes curtained in by
the darkness and again veiled by the light; its investiture is mortal and its fruition is death.

The spiritual espousal, wherein humanity is united
with the Lord, is not only catholic, including all the
elements in a human world, but, whatever may be its
heavenly consummation, is, in its earthly expression
and as a visible manifestation, a limited estate, involving conditions such as attend all other espousals : on
the Bride's part a destination separating her from the
Bridegroom, and in many ways seeming a contradiction
of her inmost desire for Him, so that she becomes a
poor starveling, a distraught and desolate Psyche, be-
reft of Love; and on the part of the Bridegroom a running after her, as if in answer to some great need and
hunger developed in her desolation, as if He had in-
dulged her aversion that He might follow her into her
darkest hiding, standing at her door and knocking while

His locks are wet with the cold dews of her night—He also having veiled His essential might and brightness lest she should be dismayed at His coming, yet retaining enough of His original majesty that she may see Him as the one altogether lovely, the wonderful.

Such, at least, is the modest human regard of this spiritual marriage, which includes and transcends all the other espousals for which the world is made: the Bride taking upon her all the blame, the reproach of her very destiny. This has been the cry of the human soul since its bondage began: Mine is the shame, the low estate; there is none good but the One. But all ways the Bridegroom answered: Fear not, in thee only is My delight; Mine is the darkness and the evil, and no glory belongs to Me that is not also thine. These are the everlasting Canticles.

This similitude of a conjugal relation between man and the Lord has been a symbol familiar to the religious thought of the race in all ages, from the Vedic Hymns to Swedenborg, and is especially frequent in Hebrew prophecy. Science shows that all cosmic life is expressed through repulsions turned into attractions and affinities — what seems repulsion being itself an undisclosed, or hidden, attraction; and if we substitute living terms for these, we see that universally Nature is the harmony of conjugal associations, in all of which the primary note seems to emphasise disjunction.

II

Now when the Bridegroom was seen as Emmanuel, in whom were manifest the power and wisdom of crea-

Continuance of the Bondage.

tive Life that had been hidden beneath veils through which now it shone; when he was healing the sick and making the blind to see and releasing the captives of every earthly bondage, it seemed then to those who witnessed these things that the bondage itself was ended. "If thou hadst been here, our brother would not have died," said the two sisters of Lazarus. And seeing in him this Life as not only curative but redemptive, men said that now there need be no more sin, since here was a living stream which turned its scarlet white.

But while they were saying this the Bridegroom said, " I must go away."

So the sun had nightly left the Earth to her darkness and yearly to her winter, since the Night and the Winter had their own work to do with the Earth—the very complement of his.

When the Lord left men to their old bondage of death and sin; left even those whom he had healed or raised from the dead to yet again sicken and die—it was evident that it was no part of his mission to abolish the captivity or to reverse the lines of development in the world or in man as connected with the world. As we have seen, the creative life manifest in him was a singular illumination of what this life had been doing in the world and in man; it was a revelation of the truth hidden in the bondage itself, and to be expressed only in its fulfilment; and since it was the life of the Father in him, it was something more than a disclosure, transforming man's view of his finite, mortal, and sinful state : before it could be this marvellous revelation, it must be creative in the human heart, making therein a kingdom, whose principle was a working power in the world, having, indeed, a worldliness of its own in a visible social organism, and at the same time making for heavenliness — for an estate native to man as the child of God and the heir to eternal life, for a kingdom not of this world. It was to be at once an earthly unfolding and a heavenly involution. It was for this involution that the Bridegroom must go away. There was a World to Come, a new habit and habitation, a transformed Bridegroom accordingly. "I go to prepare a place for you." The new expansion involved new distance. "And I, if I be lifted up, will draw all men unto me." Moreover, for the completion of his revelation he must show not only the creative life, but that Death is creative. He must die for resurrection—to give the note of the new harmony, the theme of the spiritual life.

If the Lord had remained forever, continuing his manifestation of the Creative life, annulling sickness and death and sin, as well as all natural evil, instead of unmasquing all these, then all the discords attendant upon this harmony, which is the one known to us here, would have been resolved not positively but by negation. The harmony itself would be confined within its own null perfection, with no openness to a World to Come. It would have been an arrest of creation ; nay, more, it would have been a consummate illustration of

the folly of creation itself, which eternally includes
the Evil, and in every new specialisation—in human
existence most of all—accumulating and exaggerating
the Evil.

III

There must be the full human comprehension of
Evil, like the divine comprehension, before we can understand that our inheritance of the earth is of all destinies known to us the most glorious—the ultimate expression, so far as we yet know, of the divine will and
pleasure. The sense of this is our only assurance of a more glorious world to come.
For what hope have we if the Father's work
hitherto has so far miscarried that redemption must
mean the reversal of its whole procedure in Time?
Surely we derive no help or consolation from the belief
that either fallen man or fallen angels have been able
to oppose His will with even temporary success.

The difficulty or problem is not in the divine creation, but in our partial conception of it. What seems
to us an opposition or resistance to the divine will is an
essential element in its operation. There is no reason-
ableness in the supposition that God created Evil in order that He might destroy it, or that the specialisation of
life should have its ultimate issue in a human consciousness involving not merely fallibility, but falling, as the
very condition of its progress, in order that He might
redeem man from that estate. Evil is not for the sake
of Good. While it is true that life is from death, that
good comes from evil, and that pain is a discipline, yet

*The Hidden
Glory of our
Earthly Life.*

these issues are no adequate explication of death, evil,
and pain.   Our idea of the good is as partial as that of
the evil, and the deeper our insight the more difficult it
becomes to separate the one from the other, each indeed being comprehensible only in terms of the other ;
in a vision perfectly whole Evil would be seen to be the
other name of Good.   In the series of creative specialisations the more advanced and complex existence
multiplies and emphasises all that goes under either
name, not because evil is necessary to good or good to
evil, but because the reality underlying either conception is essential and eternal—proper to Life.   Lucifer
is Light-bearer, the morning star, and whatever disguises
he may take in falling, there can be no new dawn
that shall not witness his rising in his original brightness.

Nothing can be whole, or positively holy, which does
not include evil, the negation of which would also annul
goodness.   We say that God makes the wrath of man
to praise Him : aye, and but for wrath, human and divine, there would be neither praise nor praiseworthiness.
Hate is Love's other name, as Evil is that of Good.

Christ came not to destroy or to reverse the Father's
work, but to fulfil it.

In the bewilderment of our Garden, so enclosed,
whose springs are hidden and whose fountains sealed,
where we have eaten of one tree while a sword guards
the other; where Love takes on the masques of anger and hate, emphasising division and strife ; where
pleasure begins and ends in pain ; where motion begins
in disturbance and ends in ruin ; and where the advance of life and the enhancement of its charms are

through the more and more complicate involvement of
bondage, through the multiplication of perils and solicitudes, and through a constantly increasing capacity
for the inclusion of death as well as an accumulation
outwardly of the mortal structure and fabric : in this
estate the stress and travail are conspicuous, and the
glory of our existence is hidden.   But it is only that
the Bridegroom may surprise us, shining through every
fold of our heavy vesture, lifting the clouds in our
sky, lightening our burdens, disclosing the redemptive
course of evil and unmasquing death.   To His vision
the glory of our earthly life is ever open, tempting Him
to share it (as it does the angels), and leading Him on
to His incarnation.   We look upon this glory in Him
as a divine disclosure, but it is a re-presentation to us
of our humanity, and He stands for us and falls for us,
in our image, so that we may comprehend our standing
and falling, in His image.

Redemption is the other name of creation—the luminous reflection and complement of all in creative
specialisation that we call evil.

IV

The bondage, then, is continued and completed in
the spiritual organisation which we know as Christendom, and which is the coming in all flesh
of the Kingdom whose principle was expressed in the Lord incarnate—expressed
for what it is essentially, as the principle
of an eternal life.

Human Fallibility in Christian Experience.

The Bridegroom was always visiting humanity before He came in the flesh, and always had a spiritual kingdom in human hearts. After His ascension, in a body already adumbrating that wherewith all the Children of the Resurrection shall hereafter be clothed, He was still a real presence in His earthly kingdom—a kingdom including all the evil of the world and all that belongs to man in his sinful and mortal estate. Even the regenerate, while in the flesh, retain the fallibility which humanity has had from the beginning : only it has for them its full meaning. The increase and progression of the spiritual life in all outward embodiment and development is a planetary wandering, a prodigal exile, showing often a ragged vesture, and full of repentances. The authority of this life, being one with its growth, does not exclude but depends upon the human fallibility. It is an experience.

The ecclesiastical not less than the secular history of Christendom is an illustration of fallibility as a condition of progress. The movement is a succession of nights and mornings, of stumblings and ascents. Always the aversion from the Bridegroom is followed by a fuller reception of Him. Often it seems that Christ is asleep in his disciples' bark while the storm is brewing : nevertheless, the storm is his as is the calm.

V

In the Christian world outside of the ecclesiastical system all development seems to contradict the Sermon on the Mount, and this opposition follows the

laws of life expressed in all organic structure and functioning since the world began.   Only thus

Contradiction of the System to its Principle. does Christianity maintain its existence as a
working power in human society.   The king-
dom of God on earth is an integration—not
merely an inward wholeness, as it was in the singular
destiny of the Hebrew people, but an outward organisation seeking completeness in polity, art, philosophy,
and ethics ; and the more earnestly it pursues these
lines the more it has of inward grace, vitality, and illumination.   The glory of Christianity is chiefly manifest in that it is a continually lifting and transforming
power notwithstanding its inclusion of evil, nay, by
virtue thereof, since no new ascent is made save through
descent and apparent recession.

Christian peoples accept the vital principle illuminated by Hebrew prophecy and by the life and teach-
ings of Jesus, but they do not repeat the process through
which that luminous revelation was vouchsafed to them.
Rather they appear to contradict it, seeking especially
that outward excellence and accomplishment which were
denied to the Hebrew exemplar.   Nor does the individual Christian repeat the divestiture of the Lord's
life.   He follows, but he avoids the exact similitude.
The original exemplar, bringing into clear light what
had been hidden, would have been marred and confused by that outward fabric and equipment which had
always been its obscuration.   Emphasis given to even
the outward moral habit would have disguised the light of
life.   Nevertheless, the very elements which would have
blurred the central light—which had indeed hidden it
from the beginning, and which will continue to veil it in

every earthly manifestation of it to the end—are necessary to any orderly planetary system revolving about it.

The development of the Mosaic Law obscured its
original principle. Pagan systems in like manner veiled
and in the end perverted and disguised the bright truths
which irradiated and graced their beginnings. The institutions which had so stable, so vast, and so complex
development in the Roman Empire were woven into a
fabric of conventional habit and tradition which became
dull and lifeless. Such reaction as gave them any
bright illusion came from no zeal like that of the Hebrew
prophets, but chiefly from the poets and philosophers
inspired by Greek culture ; it was not radical in reaction,
and it antagonised structural degeneration rather than
the systems themselves, whose dissolution was necessary
to any genuine renascence. The old sentiment of kinship was weakened, while the lines of caste became more
rigid ; social amenities consisted with fine cruelties ; civic
grandeur and formal justice tended to exclude living
graces, until the only really vital current was the life of
the lowly people, broken and downcast, and so prepared
to receive the Christian Gospel, while the hard, artificial
crust, lifted far above the stream, awaited the hammer
of the Goth which was to break it in pieces. Yet in all
these systems are found mundane charms, not apparent in Hebraic life, which are associated only with the
finesse of culture in manners, literature, and art, being
inseparable from a stable order of things having the fe-
icity of outward completeness, in a movement not hasti-
ly arrested by violence from without, by holy zeal, or by
prophetic paralysis, but allowed its natural modulation
and conclusion.

Because the Hebrew race, or that remnant of it which was held to its peculiar destiny, was withheld from the outward accomplishments which have constituted the greatness of other peoples, it is not therefore to be accepted as the model of national development. The little child is the type of the spiritual life of the Christian; but the Christian is not therefore denied the sturdy maturity of manhood. The ethical conception of the Greek, Roman, or modern world is not prominent in the Sermon on the Mount, but we are not therefore called upon to repudiate ethics, or even that social specialisation of morality which seems to contradict the words of the Master. We do not instruct our police to ignore the overt act and to regard only the inward motive; we maintain our conventional procedure in government and in all social functions; and in the conduct of our individual life we do not practise celibacy because the Lord did not marry; though he said, Give to him that asketh, we do not indulge ourselves in indiscriminate alms - giving, nor do we discard prudence because he said, Take no thought for the morrow.

The disintegration of Hebrew life and that divestiture which characterised the life of the Lord and his disciples served a singular purpose for all humanity, baring the inmost heart, the supreme desire, "the one thing needful." That purpose was served so effectively that the true Christian can never lose sight of the spiritual principle. While there are circumstances in which men who would secure the greatest fruitfulness of work for others must be " eunuchs for the Kingdom of Heaven's sake," freshly illustrating the central principle of their faith, yet from the foundation laid by these must be erected a superstructure which shall at the same time express the divine-human fellowship and the economies of a complex social order, civil, moral, intellectual, æsthetic, and industrial. There are times when the preacher must take the humble garb of the prophet, and, like St. Francis of Assisi, teach the lesson of poverty; and there are periods of wide-spread corruption and dead formalism, when superstructures must be destroyed, and the over-ripe and morbid summer, vaunting her distainment and reeking in wantonness, must yield to the rigour and release of winter. But for Puritan, Methodist, and Quaker—for all the prophets of divestiture—there is the spring-time also and the foison of another summer.

VI

Either season has its evils as well as its goods: its characteristic violence, whether it be the fanaticism of destruction or the madness of merrymaking; and its peculiar grace, whether it be that of candid and unyielding virtue, or that of virtue's sacrifice. Christianity frankly owns both seasons. The Lord himself, in the most sublime utterance that ever fell from human lips, said: "Whereunto then shall I liken the men of this generation? . . . They are like unto children that sit in the marketplace, and call one to another, which say, We piped unto you, and ye did not dance; we wailed, and ye did not weep. For John the Baptist is come eating no bread nor drinking wine; and ye say, He hath a devil. The Son of man is come eating and drinking; and ye say, Behold, a

The Summer and Winter of Life.

gluttonous man, and a winebibber, a friend of publicans and sinners! But wisdom is justified of all her children."

It is the bridegroom's presence that prompts the festival, and in his absence there is fasting. Devoid as was the life of Christ of everything associated with material wealth and worldly pomp, yet the seed of his kingdom, which in him suffered death, divesting itself of every outward integument, so that it was seen in the naked essence of its germinant power, was to abound in the world because of that death, showing its heavenly might in earthly investiture.

Christ as a Prophet reversed the prophet's primitive habit. Among all Oriental peoples the earliest manifestation of prophecy was attended with a kind of frenzy, with wild antics, repellent yet fascinating and awe-inspiring, like the frantic mood of a Delphic priestess. Islam began in epilepsy. Prophecy in these aspects is corrosive and like a biting frost, with an eager momentum of destruction. It tears away all veils, as does insanity, and dispels illusions. Life in its fresh vigour turns away from this hoary violence and seeks investment and plenitude, dramatic masques, the full volume of its harmony, the momentum of its procession. But this movement also comes into its fever and drastic violence.

Storage is for expenditure, and the expenditure runs into ruin, so that there seems to be the divine law of impoverishment, bringing desire back to its hunger. But the hungry are blessed only because they shall be filled. If one rests in the hunger for its own sake, then has it a greater peril than gluttony and drunkenness, as is illustrated in the temptations of St. Anthony. The empty room, swept and garnished, is especially prepared for demoniacal possession.

The Messiah did not come to men as an impalpable ghost (even after his resurrection), inviting them to disembodiment. Rather was our human flesh as dear to him as that of children to their mother, and never in word of his was there any animadversion upon our carnal plight. He enjoyed the festival, and even turned water into wine for those already well-drunken.

VII

While the deepest spiritual insight reverts to the Child Jesus and to the plasticity of the Christian type in his followers; to the love which judgeth not and thinketh no evil, yet it is a view which may be so held as to arrest all development, and to neutralise Gospel Antitheses. Christianity as an organ of social movement and as a working power in the world. The injunction to turn the other cheek also to the smiter is one that if followed would truly express the spiritual attitude of the Christian toward all men, as preferring peace to strife. But the Lord himself gave quite another view of the practical operation of Christianity as a promoter of strife, setting a man at variance with those of his own household. " The zeal of thine house hath eaten me up ;" but this zeal for the inmost Presence became in the outer court a flagellation of those who made it a den of thieves. He bade his disciples to pray in secret to Him who seeth in secret, and in alms-giving to

not let the left hand know what the right hand doeth—
as if goodness had only a hidden excellence, and should
be removed from the field of self-consciousness. Yet
he bade them let their light shine before men that these
might see their good works. The children of the household were free from obligation to Cæsar, yet he advised
the payment of the tribute. The miracle was the sign of
the hidden potency of the life that was in him, but he
exercised this power reluctantly, and declared wicked
and adulterous the generation seeking the sign. " If
they hear not Moses and the prophets, neither will they
be persuaded, if one rise from the dead."

The principle of the heavenly kingdom was flexible,
spontaneous in its operation, as of a spirit that is like
the wind, which bloweth where it listeth, and thou
canst not tell whence it cometh nor whither it goeth.
Yet in many ways the Lord recognised as necessary an
order which tends to hardness and firm stability, as of
a house founded upon a rock. The hard lines of development are not ignored. Strait is the gate, and
narrow is the way. Strive to enter in. Thou knewest
that I was a hard master, gathering where I have not
strewn; therefore even thy one talent must not be
hidden, but must return to me with usury. Seek, and ye
shall find, knock, and it shall be opened unto you. By
their fruits ye shall know the Children of the Father.

Christ himself had come into an order more ancient
than the earth; he had always been in it, the creative
life thereof, determining its course; but he had now
come into it as man, with all the passions of a man,
with all the limitations of a human consciousness; and
he had come into it not for its abrogation but for its

fulfilment. Christianity was in its organisation to be the fulfilment for man of his destiny in the course already begun, including all human elements ; it was to be an order as an organised human experience. With the harmlessness of the dove was to be united the wisdom of the serpent—that very wisdom which led man out of Eden.

The divine temptation leads us into the illusions of the phenomenal world. The divine redemption participates in these illusions. The coming of the Lord was an appearing. But he made all veils transparent.

VIII

Any religious system which should profess to rend all veils, which should attempt the abrogation of time and the world, and of the desire which makes its way outwardly into worldly embodiments *The two Extremes of Religious Systems.* and constructions, would rest in Buddhistic nihilism. This is "the will not to live," the characteristic, or rather the characterless, aim of Schopenhauer's pessimistic philosophy. It is not one with the divine will, and it is not an acceptance or comprehension of that will, but is rather its repudiation.

Mahometanism, going to the other extreme, even promising to its adherents a sensual Paradise, frankly accepts all illusions, but makes them the everlasting cerements of the soul. Islam is the modern Ishmael, whose hand is against every man, and every man's hand against him. This faith began in the insanity of the prophetic function unaccompanied by prophetic insight;

began in brutalities, and has progressed through conquest based upon insolence and signalised by its atrocities. It is perversely dissociative, incapable of catholic fellowship, or even of coherence among its own constituents. It has been of service to the modern world chiefly as a menace and a challenge, holding shrines not its own, and so provoking the crusades, and promoting organisation as against itself, very much as Napoleon caused the rehabilitation of Europe as the sole means of its security against his inordinate rapacity.

Christendom, mainly Indo-European in its constitution ; anti-Semitic, though deriving its religious inspiration from the Hebrew ; in its westward course of empire never wholly losing its inward orientation, has been allowed its steady growth because the monstrous aggregations of humanity in Asia have slumbered, the Moslem alone having shown a strong hand, but disturbing only to stimulate.

IX

The Christianity which has made this Christendom did not owe its first expansion in the West to its organisation. It was in its plastic childhood, and when it seemed most averse to worldly offices and emoluments, when its ritual was a simple, homely affair not yet associated with Church edifices, that it established its contacts with the world, spreading as a gentle insinuation throughout the Roman empire. This was also the time of its inspired writings. The wonderful expansion and inspiration were miracles such as belong to infancy, spontaneous manifestations

Primitive Christianity.

native to the spirit and not apparent, but rather obscured
in later periods of structural development. A mighty
wave of heavenly strength and peace seemed to pass
over the whole earth, quite in accord with the conditions of the general armistice then prevailing, and especially comfortable to the down-trodden and distressed
poor, to whom no worldly armistice brought rest or consolation. The Gospels and the Epistles breathed the
spirit of love and peace, bidding men love one another
and bear each other's burdens. At Jerusalem, where
there was the greatest tenacity of the old forms, and
also the insistence upon justification by works, before the
name of Christian was adopted by the Church, the followers of Jesus in a singular manner illustrated the gracious spirit of a new faith in a communistic economy.
It was a mode of life that could not be maintained, and
the Christians at Jerusalem became, in consequence of
it, a burden upon the Western churches; but it lasted
long enough to find expression in the sublime ethics of
St. James's Epistle, which shows what the moral order
may become when wholly vitalised by the spirit of
Christ, and when society, though it may not have become communistic, shall in its economic expression
have reconciled with the law of Love all the competitions and antagonisms necessary to outward integration
and development. This reconcilement is indicated in
those words of the Lord, not recorded in the gospels,
but quoted in early Christian writings: "When the outside is as the inside, then the kingdom of heaven is
come." Only in that consummation visibly realised
could we see what was the scope of the kingdom determined in its marvellous germination. The definite an-

ticipation of the issue is not possible even in the most
hopeful dream of the optimist.

In a very vital sense there *was* organisation even in
this earliest manifestation of Christianity. The fellowship was itself a living organism, a vine with tender
branches widely and swiftly spreading, throwing its soft
tendrils about the hearts of the lowly, and thus for a
long time escaping the notice of the powerful. This
was its native disposition, following closely the ways of
the Master. We think of it as a power building up
from the bottom, and so it was, if we consider only its
main constituency ; but in a society like that of the
pagan world at this period—a world prepared for its
own dissolution, and expecting, as in a dream, some
transformation from a mysterious source—there are always wise men, and wise especially in the culture of the
heart, to whom nothing human is alien ; who for humanity are willing to give up class, in an order where no
class is fortunate and all are at a loss ; who are looking
for some new star of hope in their heavens. To such
men Christianity from the first made a strong appeal,
and they naturally became the leaders of the people.
Others there may have been, men of religious zeal and
high intellectual attainments, who, like Saul of Tarsus,
first came into contact with the Christians as their per-
secutors, and, seeing in the new faith a greater motive
for their zeal, became its ardent adherents.

X

Certainly the ecclesiastical organisation must have been far advanced, and must have shown a disposition toward authority and influence in society and the State, when Constantine became the champion of Christianity, and took its symbol of the cross as the sign through which his armies should become victorious.

When at a later period the Church came into close alliance with the State, becoming the arbiter of empires, its organisation as a world-power had complete development, entering into the full amplitude of its earthly investiture. Catholic *The Mediæ-* *val Church.* brotherly love was at the heart of it, and in every fold of its garment. It was the cosmic order of the Lord's spiritual kingdom — the field of the Lord's espousal with humanity. That was a true pontificate which bridged all the chasms between social classes — between wealth and poverty, culture and ignorance, mastery and service, and also between heavenly grace and the arbitrary limitations of formal justice. It was such a hierarchy as naturally found its typical representative in St. Augustine.

The Church had placed in the hands of the Roman Pontiff not only the crosier, but also the sword and sceptre ; and the social order of Christendom in the mediæval period could not otherwise have been established and maintained on a Christian basis. Not less but more than in the age of primitive Christianity was this organisation the embodiment of the Spirit, for, though the pentecostal flame was hidden, yet it was the same

flame that vitalised the whole structure in its vigorous
growth for the full measure of its beneficent ministration. The dove, which is the emblem of the Holy Spirit,
because it has wings for flight, does not therefore make
his fixed abode in the heavens, but rather descends and
makes his home among the haunts of men. The ministration of the Spirit is by descent. It was so in the
Christ; it is so in the Church, which as a fellowship is
everlasting, but which as visibly manifested in any special embodiment has a beneficence in its expenditure
and even in its disintegration measured by the degree
in which it has received and manifested the spirit of
fellowship. As a world-power an ecclesiastical, like any
other organism, rises to the height from which it may
most beneficently fall.

That alliance of Church and State in which the former
was authoritatively dominant lasted long enough to secure its ends; and during this period the wisdom of the
serpent, so necessary to its efficiency, was fully evident
in its practical working and in its development of dogma. The exigencies of the ecclesiastical situation de-
manded a dramatic theology as well as a dramatic
ritual; and in both became manifest the inevitable contradiction of the formal system to its formative principle.

Regarding merely external appearances, it would seem
that the integration of Christendom had been secured
by the surrender of Christianity itself. The Church
would appear to have been dominated by the world.
The Protestant reformers easily substituted for the scarlet woman of the Apocalypse, there indicating the Rome
of Nero, the papal Rome of their own century. But in

reality an inestimable service had been rendered to humanity by the mediæval Church. Pagan Europe had been brought into the Christian fold ; among the common people the faith had been accepted in its simplicity, and, though mingled with superstitious imaginings, it had nourished and brought into activity the sentiments and impulses peculiarly distinctive to a Christian life, individual and social. The people were lifted into a freer atmosphere and yet remained unsophisticated, readily moved by generous enthusiasms and hospitable to the lofty motives of an age which abounded in chivalric romance and saintly legend. They interpreted so much of the Gospel as reached them with their hearts rather than with their intellects. Theology and ritual were the concern of the bishops, and the side of these presented to the popular heart was that best ministering to its need—impressive, nutritive, and disciplinary. Thus was preserved within the hard enclosure of official ecclesiasticism a genuine spiritual fellowship ; for this indeed was the induration of the system necessary, as government is necessary for the protection of home life and social activities. No system retaining the simple plasticity of primitive Christianity could have withstood the invasion of Islam ; nor would it have sufficed for the building up of Christendom through the tutelage and discipline of the swarming Barbarians whose rude strength had throttled Roman civilisation.

The degeneration and corruption which in a natural sequence followed this ecclesiastical evolution were but the accidents attending the completion of a sacrifice begun in the fortitude of a necessary but arbitrary sovereignty; and the forces of the Reformation were nour-

ished by the fortitude and found their opportunity in
the weakness and corruption of a structure which had
done its work in the world.

XI

The popular life in the Middle Ages owed to the
Church its happiest moods, and the natural and spontaneous exaltation of these.   The plastic state of childhood was marvellously maintained.   Faith
was creative, the builder of cathedrals, the
maker of legends ; and, as in the creation of
the world, it included the grotesque as well as the beautiful.   As the child fondles fear and insists upon the
dragon element in the fairy-tale, naïvely clinging to the
"mark of the beast" in every fanciful representation,
so the mediæval Christian imagination, with the divine
catholicity which saw the original creation to be good—
though including radical evil and all dark provisions—
freely mingled old Titanic glooms with new-born hopes,
cherishing the fiery baptism of purgatorial pains.   In
the creations of art, the ugly and miscreant had their
place in the triumphant harmony.   All things were together "bound under hope."   As the child expects the
loving spell that shall show the Beast to be really beautiful beneath his unshapely masque, so Christian love
judgeth not, but awaits that vision whose light shall
eclipse discrimination between the clean and unclean of
God's creatures, showing what we call ugly really beautiful after a pattern older than we see in what appears
to us most comely.

*Accommoda-
tion to the
Popular Life.*

Certain indulgences and accommodations of the mediæval Church to the popular mood, both in the matter
of belief and practice, seem quite natural from this point
of view : such, for example, as the tolerance of Mariolatry
among peoples accustomed to the worship of Isis and
other female divinities, and the adoption of pagan
feast-days. The growth of the kingdom was from a
seed that might be planted in any human heart, just
where that heart was found, sure to burst its cerements
and to find its proper nutriment even in the husks of
an outworn faith—to shed the false and rise the true.

Nurture itself implies a life diminished and broken
for the increase and integrity of the nursling, so that
Christian beliefs have often had in outward form the
fallibility peculiar to the estate of humanity : not corrupt or corrupting as received by the fervid believer,
though if not thus hungrily taken into his organic spiritual life, if regarded as having a use and meaning
apart from such spiritual assimilation, or if received by
the mind only as logical formulations, they, like the un-
consumed manna in the wilderness, disclose their corruptibility. The " means of grace " are not objects of
worship ; it is some descent in them from the heavenly
height of the principle they embody which brings them
next the craving of a spiritual hunger, and but for the
expedition of that satisfaction they suffer vilification.

It is not a matter of indifference what a man believes,
or what otherwise may be offered for his spiritual nourishment. The same food is not suited to all physical
organisms, or to any one organism at every stage of its
growth. The kingdom of heaven is within us, and hence
there is in us its spiritual hunger, which determines its

own selection ; and because of the marvellous growth
of this kingdom, there is a development of the hunger
itself and also of the nurture—the source and principle,
in either case remaining the same, being essential and
eternal. But the growth is an ascension, and that which
ministers to it a descension. This is the ministration
of death unto life.

<h1 style="text-align:center">XII</h1>

Protestantism, however, was very far from being a
revival of primitive Christianity.   Luther, indeed, revived Paul's doctrine of Justification by Faith, and

*Ecclesias-
tical Special-
isation.*

with such vehemence that he denounced
the Epistle of James as " an epistle of straw ;"
but the movement of the Reformation was
itself so far dominated by State policy that its immediate result seemed to be a mere schism rather than the
great spiritual reaction which radically it really was.
Wherever this reaction was not at first evident it was
afterward fully developed, as in English Puritanism.

The dissension itself, like that which originally had
divided the Western from the Eastern Church, was for
new integration, and it was attended with violence and
persecutive hate, such as in the Athanasian Creed had
consigned to eternal damnation all Christians not as-
senting to its doctrine—showing that not only the wisdom of the serpent but its venom also entered into the
ecclesiastical edification, even as the horrors of war
mark every critical epoch in the progress of civilisation.
There is no nutritive process, for the building up of any
structure, that does not involve the production of poison,

and still more conspicuously does this malady attend all organic functioning.

Christian fellowship does not, even in its beginning, mean the destruction of antipathies, and the divine life no more than the human has for its aim security, peace, and quietness. That would be to substitute salvation for redemption. Ecclesiastical, like all other specialisation, is through division. It is as inevitable that the visible Church should be broken up into sects as that a vast empire should be divided between different races— each of these developing a separate nationality. This tendency leads, as disciplined intelligence becomes general, to individualism and the emphatic recognition of personal liberty and responsibility.

Our Christian civilisation is fortunate in having reached a point, never even approached by any ancient civilisation, where we can frankly give up the poet's dream of

" The Parliament of man, the Federation of the world."

The individual does not wither as the world grows more and more. He who in the true sense is most himself is most for the world. The profoundest patriotism is the truest cosmopolitanism. We can already see that the kingdom of heaven cometh not by observation. It is no external dynastic bond that can unite nations : the outward delimitation promotes the inward bond. It is fortunate for both State and Church that the social order has entered upon that stage in its progression in which each can best perform its functions independently of the other, and in such manner as to leave the individual, in his proper field, perfectly free, unconscious of any outward authority exercised by either ; fortunate also for society that it can hope in the near future to have the perfectly free play of all its proper activities in the development of industry, science, and art.

This is, indeed, the sum of the advance made by Christendom since the Renaissance, which gave to the modern world all that was worth having from the old— not as a mere heritage, but as something to be crea- tively transformed by the Christian spirit.

In all these lines of advance the kingdom of heaven after the Christ type has its specialisation. It is the specialisation of humanity—not of a visible Church, of a visible State, or of that which we call Society ; least of all is it a realisation of St. Augustine's *Civitas Dei :* all these are but the masques of the surely though invisibly coming kingdom. Other masques will follow these, the same veils indeed, but clarified and made transparent in the process of human redemption. The dramatic theology of St. Augustine, so alien to the conceptions of redemption entertained by Paul and by the Greek Fa- thers, with its peculiar doctrine of Grace as confined within sacramental limitations, must, like the dramatic pomp of ritual, pass into its drastic stage and disappear. The meanings of the divine Logos, as manifested in Nature and in the Incarnation, will be ultimately as they were primarily seen to be for all humanity, and to themselves transcend an historical Christ, an Apostolic succession, and a limited fellowship.

XIII

It seems strange that at the very stage of progression, when this noble prospect is possible, the superficial view of our civilisation is made the basis of the profoundest pessimism. But it is in this very field of pessimism that the Christian finds the signs of his brightest hope. In his view the rigid worldly mechanism becomes celestial, and materialism is seen as solvent to the Spirit of Life. The automatism of habit, a facile descent into oblivion, from which life and meaning are withdrawn, is seen as a release of life for new initiation; and though in this mortal habit the whole world should slip away it would be for the resurgence of a new world. Stability itself is kinetic, the resultant of velocities inconceivably swift. The diabolism, which in the old systems of dualism was regarded as inherent in matter, is exorcised.

Ready Reaction of Modern Life.

The Christian idea of Death, confirmed by every disclosure of science, is itself that of solution, through the reaction proper to Life. The Christian idea of a universal human fellowship, a recognition of the eternal kinship, gives to Christendom its scope, broad enough to include all reactions in the harmonious interaction of all the forces and elements involved. The fundamental difference between Paganism and Christendom is that the latter, though its systems fail, has within itself the secret principle of renascence, so that the Child Jesus is forever being born. Owing to the readiness of reaction, which increases with the expansion of knowledge among all classes of the people, the Order, like a living organism, is conserved through its inclusion of death, and revolutions are possible without that extreme violence which marked those of earlier times. That in the system which falls is doing its work.

All specialisation is a hiding of Life, whose authority in our human progression is thus secluded from the authority of institutions, retaining its creative potency. At every step in advance something is given up which to our backward look seems more precious than what we have gained. Thus we regret the picturesque mediæval life with its marvellous enthusiasms, its chivalric impulse, and romantic heroism, even as many souls in that period regretted paganism and longed for the return of Pan. Even in the emancipation of our slaves we seem to have suffered a loss through the rupture of an intimate bond of affection like that which holds together the members of a household.

The gain from these successive revulsions is apparent from a wider view. Every emancipation is an entrance upon a life involving severer limitations, but the enlargement of our perspective and the free play of our emotional and intellectual activities depend upon this complexity of our finitude. In the discreteness of the special accord is its proper excellence and also its correspondence to the universal harmony. The complete perspective would receive the full pulsation of the eternal life and its full illumination.

The Hidden Life—our life hidden with Christ in God—is our eternal and inalienable heritage. The issues of this life in the visible world, in the procession of generations, we cannot mentally anticipate, nor are they disclosed in any prophecy. The creative specialisation will go on, and will surely be completed in redemption. Action will still be reaction, antipathy resolved as sympathy, repulsion as attraction, bondage as freedom, and death as swallowed up of life. Evil—all that we have called evil from the beginning — will remain, even as darkness will alternate with light, and to whatever extent abnormal perversion, inordinate selfishness, and arbitrary caprice—the accidents of a partially completed order—may disappear, life will still have its normal pathology—its pain and frailty and repentance.

CHAPTER IV

ANOTHER WORLD

WHAT do we or can we know about the thither side of Death?

There is no sequel to the story of Lazarus, who was raised from the dead, disclosing the secrets of that estate which had been a reality to him for four days, as we count time upon the earth.

The Lord himself, the revealer, in a singular sense, of spiritual truth, and especially the illuminator of Death, gave, so far as we know, no intimation to his disciples of the life beyond the grave. Nor is it recorded that they asked for any. Death was unmasqued in the Resurrection and was shown as one with creation, but the full light of this wonderful illumination was thrown upon life here, showing not one definite lineament, not even a shadowy trace of the life beyond. There never has been any but an imaginative disclosure of that life to men living upon the earth.

A curtain drawn so closely about the present existence must have excited the vivid curiosity of the pagan mind. We find in ancient literature no trace of this curiosity in the shape it takes in recent times, because it was so vivid and therefore so immediately took a fixed shape in an imagination whose constructions were real beyond the shadow of a doubt. There was a development of this imagination from age to age, but at every point its creations were regarded as unquestionable realities, as certain as the objects of present experience. To the Egyptian the *Book of the Dead* was a genuine and trustworthy itinerary. What Polygnotus painted or Homer described concerning Hades was but ˗ rescript of what the Greek already knew with unwavering assurance.

It was only when, in a comparatively recent period, men began to question the reality of the immediately external world and to impugn the trustworthiness of their senses that the "other world" also became unstable and the sport of a mutable fancy.

When we say that the ancient imaginations of the unseen world were held as certitudes, like the sense-perceptions of objects in the visible world, it is not therefore to be supposed that these imaginations constituted a real knowledge. Indeed, our sense-perceptions do not constitute a real knowledge of the external world with which we are in contact; how much less truly could imagination render to us the world beyond. The belief which men have had in such imaginings is very much like our belief in dreams which seem to us real even though, in some deeper consciousness, we know that we are dreaming. Any disclosure or communication must be in the terms of a life that now is; and sensibility—whatever illusion it may involve—is at least this vital and present contact. But men have always suspected the masque of the world in their sensibility. It is not likely that the more complex disguise of the imagination has at any time escaped suspicion. In the background of all human thinking, however crude, has been this intuition : we know only that which, knowing, we do not know that we know. Gnosticism is of the eternal. Conscious knowledge is of things in time— present, past, and future—things veiled by virtue of manifestation. "Another world," considered as a definite existence, is the only field for absolute agnosticism, wholly cut off from human knowledge through sense, intellect, or spiritual apprehension ; it is not veiled but absolutely hidden, and of it there is no possible revelation, save through entrance upon its actualities, when it ceases to be "another." We know the divine, the eternal ; indeed, these alone are really known since life itself is essentially these ; but what we call another world is not simply invisible, not simply a future or a next world in the sense that we think of to-morrow or next year ; it is another by an inconceivable diversity— a distinct harmonic synthesis, for us unrelated, and untranslatable in any terms known to us. The world to come we know, since it is that which *this* world becomes. Another world is a new becoming, having its own "world to come ;" it is the only incommunicable.

No divine revelation has ever attempted to broach the inviolable secret. Eye hath not seen, ear hath not heard, neither hath it entered into the heart of man to conceive.

There is one utterance by the Lord, recorded in the Gospel, concerning the state of the Children of the Resurrection : "They shall not marry, nor be given in marriage : neither shall they die any more." It is remarkable that, in this declaration, sex and death are joined together, as science shows them to be in the specialisation of organic life.

The Lord referred to sex and death as we know them, in their specialisation. While the essential principle of espousal and that of death are eternal, proper to any life here or hereafter, it is possible to conceive of a state of existence wherein the manifestation of these involves none of the external features associated with our knowledge of them in their earthly manifestation. As there are lower organisms which we know to be sexless and deathless, in the sense we have of sex and death in an advanced specialisation, so there may be higher organisms, belonging to that "other world," to which these special terms are inapplicable. We say there may be : Christ says there are ; and although this assertion is the only one made by him directly bearing upon the conditions of a future life, it is very far-reaching in its suggestions.

Even in this earthly human. life all desire is spiritually lifted into its heaven, not as being destroyed, but as dying to one environment and being raised into another, where its manifestation takes higher forms and its ministrations seem like those of the angels. It is as if out of the earthly matrix of Passion had been born its heavenly embodiment, not associated with corruption and so seeming something deathless, though it lives through the quickness of what Death essentially is in an eternal life. It is possible that the Lord's saying had its real meaning as applicable to the heavenly exaltation of any life, present or future. Certainly the characteristic of Christian life is its realisation here of an eternal life, through a constant death and resurrection ; and this exaltation belongs to our antipathies as well as to our sympathies—to hate and anger as well as to love : these also having their heaven and angelic scope, in a field of reconcilement.

We can see, then, why Christian thought is fixed upon a World to Come rather than upon what is called Another World. This present life has part in the eternal as truly as any life ever can have.

We pass from glory to glory, and that crisis which we call death is only a transition from one harmony to another. In certain forms of the Polish national dances, the guests move from room to room in the palace, the music and the movement ever changing in the processional march, according to the progressive phases of the theme enacted. From beginning to end it is the same theme, and the guests are the same. So it may be in the progression of our human life from one mansion to another of the Father's House; there is a mystic change, not of personalities but of special individual guises, involving complete divestiture, the theme enacted remaining the same.

It is because of the complete divestiture that entire newness is possible. Our attention is so fixed upon structure and upon changes as themselves structural that we seem at a loss when the entire structure disappears from our view. But how does a structure begin? Is not birth as much a mystery as death? Form is of the essence; and, in a sense not to be expressed in language, the personality has eternal form.

> " Eternal form shall still divide
> The eternal soul from all beside."

In the same sense, familiarity in time has its ground

in the eternal familiarity, whereby alone we know and are known. Our cognition here is re-cognition.

The formed memory and the formed character may be destroyed; but the life withdrawn from these, their essential ground, has its spiritual embodiment after its distinct type, still remembering and re-cognisant. The "deeds done in the body" are not, but the doer *is*, and according to those deeds: in essential form accordant, whatever the new environment. The child seems an entirely new creature, but, whatever science may determine as to his inheritance of characteristics acquired in preceding generations, he is surely and wholly an heir in that he can himself acquire anything—an heir, not simply because of and in relation to an outward heritage, but because of what he is. There is in this continuity an inscrutable mystery: that which determines the accord in the series is invisible. It is the mystery of Genesis itself. The continuity phenomenally is through discontinuity; death is essential in birth as in growth. Now, let the break—that interval in the harmony which we call death—be, to all appearance, absolute; then the resurgence, beyond our vision, is in the very field of creation; passing out of the known series, out of the succession of what we know as in Time, it is the property of life as eternal, the heritage of the eternal kinship, *under a new limitation*.

What is the continuity from the limitation known to us to that new and wholly unimaginable limitation? The mystery is transferred from the visible to an invisible death, which is one with the invisible birth. But the new birth—what is its matrix?

Suppose we were permitted to resume a position at

a point in time before the appearance of organic life upon the earth. Would any then existing form of inorganic life help us to an imagination of physiological embodiment? Science confesses its inability to answer the question, What was the matrix of cell-life?

An equally insoluble mystery is presented, if we inquire what is the matrix of any form, or how the continuity of either a generic or an. individual type of organic life is maintained in all permutations of environment. It is a mystery belonging to creation, incommunicable, itself the ground of communication. No considerations derived from what we know of the constitution of matter or of material structures, and none derived from mental categories, explain the transformations of the visible world: how much less can they be expected to even suggest the forms and limitations of an order of existence not yet creatively communicated!

Because we, in our present existence, have no conscious knowledge of pre-existent states, it does not follow that the future life will be wholly denied such knowledge. Our conscious intelligence here is a distinctive characteristic of the ultimate order in the known series; and in man this intelligence involves peculiar powers of reflection, co-ordination, and interpretation, so that the psychical as well as the physical man surmounts the entire series resumed in him. In a new order it may be a characteristic of the creative communication that conscious intelligence shall be a clearer resumption, involving at least the conscious recognition of friends and kindred. Our cognition here of anything is unconsciously re-cognition, a seeing as through a glass darkly, a mere adumbration of a

recognition hereafter which shall be a seeing face to
face. Illusions there may be—the face itself is a veil
—but there may be a more transparent mediation in
the communication, undisturbed by the obscurations
and refractions such as limit our present mental vision.
We speak of what *may* be; every presumption of a
revelation which is itself a transcendent creative communication gives assurance instead of mere hypothesis.

To our reason this subject is beset with difficulties,
because we become entangled in dilemmas suggested
by present relations, such as imprisoned the minds of
the Sadducees in the problem they presented to Christ.

Because the new assumption or embodiment is not
of flesh and blood, as we know them, it is not necessary
to suppose that it is immaterial. To it a new sensibility and a new thought would involve space and time as
forms to which our corresponding terms for these would
be merely analogues.

Given us a new sensibility, there would be given us a
new universe. We say the dead have passed away from
us, but it is perfectly reasonable to conceive of them as
nearer to us than ever, in a closer intimacy than any
known to us.

During the century now closing man has made an
important advance through dealing with subtle cosmic
forces which had hitherto been known only as dealing
with him, and, even thus, scarcely appreciated. Electrical phenomena had been observed in sparks occa-
sioned by friction and in the lightning, and the magnetic
current had been utilised in the compass ; but the terms
electricity and magnetism had but a glint of the mean-

ing now attached to them. We do not yet know what these invisible currents are, but we have made ourselves at home with them, and comprehend what formerly was not suspected—their intimacies with all cosmic operation and with our animate economies. For the obvious terrestrial forces, manifest in weight and pressure and elasticity, we are now rapidly substituting these finer tensions, thus driving the horses of the sun without risking the fate of Icarus. It is as if our solar heritage had been restored to us. Through this widened familiarity in a field which until so recent a period was wholly hidden from us, we have reached a new and etherealised conception of matter, and have come to feel the pulse of a living universe. Science is redeeming matter, making its veils transparent.

In this new view it is not difficult for us to conceive of spiritual intimacies more subtle and pervasive than any which science has disclosed in the material world, though these cannot be apparent to us in a definitely conscious appreciation.

If on the same wire, through electrical vibrations in musical accord, several distinct messages may be simultaneously conveyed, why may not all that we call matter be at the same time the medium for the expression of distinct orders of intelligences?

All reasoning proceeds through analogy, but we must be on our guard against the fallacy involved in the process. The truth in physics or chemistry can become a biological truth only by such transformation as is involved in the inorganic world becoming the organic. Any conception of our present conditions carried forward into our imagination of those pertinent to a future life must undergo an inconceivable and, to us here, impossible transformation.

What we know as good and evil, life and death, is but the analogue to these as we shall know them in another harmony. It is sufficient for us that in the Christ-life Death and Evil are unmasqued for us and reconciled with the Eternal Life. Our faith is in the Resurrection through the power of this eternal life : in what form we know not, but we know in what similitude—in the likeness of the Son of God.

For the lifting and illumination of our life here is the great disclosure made. Our Lord's resurrection brought him back to us, as if born to us a second time, showing us the nativity of a spiritual body. His new words to his disciples, instead of intimating the joys and pains of another world, dwelt upon the sufferings of the son of man before he could enter into his glory. So does our faith comprehend our travail and sorrow, finding in these the true way of life and that there is no other way. Christian philosophy, like science, finds in that which is the ground of heaviness the charm of levitation, the attraction which binds together a universe.

INDEX

ABRAHAM, 261-3.

Abstraction, 42.

Accords: Desire, in the line of special, 144; true to the original key, whatever dissonance in the procession, 157; discrete, sustained, 185; diversification of, in the organic harmony, 193; special for each new form of existence, 205, 316.

Achilles, in Hades, 44; among the maidens, a type of juvenescence, 210.

Age, wakefulness of, 19, 218.

Alternativity, 13-16.

Altruism, illustrated in all cosmic development, 112; in every economy of animal and social life, 159; excess of, in human relations, 172; Christianity substitutes identification for, 284.

Ancestor worship, 29.

Animal life, ascension of, 115.

Annihilation, virtue of, 46.

Another world, 318-27.

Antipathy becomes sympathy, from which it springs, 159.

Antitheses of the Gospel, 301-2.

Apollo, 37.

Appearance disguises Reality, 88, 140, 319.

Arbitrary, the, in human conduct, 136, 140.

Art, beginning of representative, 43.

Ascent of Life, 183.

Association from dissociation, 126, 150, 158.

Athene Parthenos, the type of outward completeness, 152.

Atoms, 186.

Attraction and repulsion, complementary, 144, 158, 236, 289.

Augustine, his mission, 307-9, 313.

Authority, genetic, 142; associated with growth, 253; grounded in fallibility, 295.

Aversion, first manifestation of Desire, 205.

BAPTISM, a burial with Christ, 280.

Barrenness, for life, 188, 287; Hebrew stress upon, 226.

Becoming involves fitness, 136.

Birth, a flight, 69; lies next to Death, 73, 184; a break with the Eternal, 143; a mystery as profound as death, 322.

Blood in Hebrew symbolism, 253.

Bridegroom, the, 40, 54, 164, 286-9, 294-5, 300.

Brotherhood, universal, 51, 62.

Buddhism, nihilistic, 303.

CELL, gospel of the, 102.

Chance, divine, 156.

Chemical adumbration of physiology, 107.

Childhood, familiarity of, with the invisible, 18; rapid investiture of, in modern life, 30; plasticity of, 203; pains of, 204; exaltation of, a withdrawal from the world and an imperative absorption, 208; *hauteur* of, 208; tension and storage of, 213; maintained into maturity, 218; unmoral, 236, 270; type of the kingdom of heaven, 237-243; in the Christ-life, 238; in the Hebrew, 240; in primitive Christianity, 304-5; in Christian peoples of the Middle Ages, 310.

Choice, 127, 135, 140, 152.

Christ, one with Nature, 54; became Sin and glorified Death, 233; falling

THE END

An Interpretation. By HENRY MILLS ALDEN. Book I. From the Beginning. Book II. The Incarnation. Book III. The Divine-Human Fellowship. pp. xli., 270. Post 8vo, Cloth, Uncut Edges and Gilt Top, $1 25; White and Gold, $2 00.

The book is a remarkable contribution to current religious literature. The author has brought to bear on the questions he discusses a wide and thorough knowledge not only of the questions themselves, but of many other lines of thought which are intimately related to them. . . . Out of all the jarring religious creeds and speculations that have marked the history of the race he deftly constructs a many-colored but harmonious flood of mosaics upon which the Son of God may walk in the mighty Temple of Eternal Truth. . . . In these days, when bald materialism has gained such a foothold even in the Christian Church, it is a hopeful sign to find a book like this, so full of genuine spirituality and yet so free from pious vapidity and cant.—*N. Y. Tribune.*

A remarkable book. The temper in which it is written is so fine, its tone is so authoritative without the semblance of dogmatism, and the sweep of thought is so large and steady that one is fain to receive it as what it claims to be, an interpretation, and so, in the radical sense of the word, a prophecy. Like prophecy in its most universal type, it is revolutionary in spirit, in obedience to an eternal conservatism; and it is only as one moves on through the phases of the evolutionary thought of the book that he fails to be startled by the quiet conclusions with which the author confronts him.—*Atlantic Monthly.*

A very notable book. . . . It gives, often in rhetoric as splendid as it is simple, the sum of all philosophy and of all theology, the revelation through nature and that in human words. Many passages are true prose poems.— *Brooklyn Eagle.*

(OVER.)

A book like this is not made in a day, and will not perish in a day. It has a mission greater than that of any modern religious work that we know of, and it would not be surprising to see it attain a hold upon the humbled human heart and the struggling human intelligence something like that of the "Imitation of Christ."—*Philadelphia Inquirer.*

It is profoundly suggestive, and on new lines and with freshness and power discusses world-old problems.—*Christian Intelligencer,* N. Y.

This book, if we mistake not, has a work to perform in the spiritual field not unworthy to be compared with that which "Ecce Homo" wrought in the sphere of practical Christianity. . . . It is such a view of human activities as this that the Christian world needs to-day. It is to the Christian, to the devout believer, to him whose spiritual faculty has been to some degree exercised, that the message of this book comes.—*Evangelist,* N. Y.

It cannot fail to increase the spirit of religion and faith in God's goodness.—*Boston Journal.*

It is this quality of individual vision and certainty which makes people listen to the new prophet who speaks in the pages of "God in His World." . . . After reading the book one puts it down with the feeling that a sensible, trustworthy person has said, with the certainty of absolute conviction, rooted in an unspeakable knowledge, "I know that these things I tell you are true."—*Boston Transcript.*

There is the throbbing heart of a living faith in this remarkable little volume.—*Philadelphia Press.*

A pleasing and thoughtful writing, clear of all dogmatism, and appealing to the highest and noblest in the human soul. The greatest questions that have ever been propounded to the human mind are here traversed in the light of reason and art and science and history.—*Chicago Inter-Ocean.*

PUBLISHED BY HARPER & BROTHERS, NEW YORK

☞ *The above work is for sale by all booksellers, or will be mailed by the publishers, postage prepaid, on receipt of price.*

www.ingramcontent.com/pod-product-compliance
Lightning Source LLC
Chambersburg PA
CBHW021113270326
41929CB00009B/863